Lecture Notes in Computational Vision and Biomechanics

Volume 29

The research related to the analysis of living structures (Biomechanics) has been a source of recent research in several distinct areas of science, for example, Mathematics, Mechanical Engineering, Physics, Informatics, Medicine and Sport. However, for its successful achievement, numerous research topics should be considered, such as image processing and analysis, geometric and numerical modelling, biomechanics, experimental analysis, mechanobiology and enhanced visualization, and their application to real cases must be developed and more investigation is needed. Additionally, enhanced hardware solutions and less invasive devices are demanded.

On the other hand, Image Analysis (Computational Vision) is used for the extraction of high level information from static images or dynamic image sequences. Examples of applications involving image analysis can be the study of motion of structures from image sequences, shape reconstruction from images, and medical diagnosis. As a multidisciplinary area, Computational Vision considers techniques and methods from other disciplines, such as Artificial Intelligence, Signal Processing, Mathematics, Physics and Informatics. Despite the many research projects in this area, more robust and efficient methods of Computational Imaging are still demanded in many application domains in Medicine, and their validation in real scenarios is matter of urgency.

These two important and predominant branches of Science are increasingly considered to be strongly connected and related. Hence, the main goal of the LNCV&B book series consists of the provision of a comprehensive forum for discussion on the current state-of-the-art in these fields by emphasizing their connection. The book series covers (but is not limited to):

- Applications of Computational Vision and Biomechanics
- Biometrics and Biomedical Pattern Analysis
- Cellular Imaging and Cellular Mechanics
- Clinical Biomechanics
- Computational Bioimaging and Visualization
- Computational Biology in Biomedical Imaging
- Development of Biomechanical Devices
- Device and Technique Development for Biomedical Imaging
- Digital Geometry Algorithms for Computational Vision and Visualization
- Experimental Biomechanics
- Gait & Posture Mechanics
- Multiscale Analysis in Biomechanics
- Neuromuscular Biomechanics
- Numerical Methods for Living Tissues
- Numerical Simulation
- Software Development on Computational Vision and Biomechanics
- Grid and High Performance Computing for Computational Vision and Biomechanics
- Image-based Geometric Modeling and Mesh Generation
- Image Processing and Analysis
- Image Processing and Visualization in Biofluids
- Image Understanding
- Material Models
- Mechanobiology
- Medical Image Analysis
- Molecular Mechanics
- Multi-Modal Image Systems
- Multiscale Biosensors in Biomedical Imaging
- Multiscale Devices and Biomems for Biomedical Imaging
- Musculoskeletal Biomechanics
- Sport Biomechanics
- Virtual Reality in Biomechanics
- Vision Systems

More information about this series at http://www.springer.com/series/8910

Sofia Brandão · Thuane Da Roza
Isabel Ramos · Teresa Mascarenhas
Editors

Women's Health and Biomechanics

Where Medicine and Engineering Meet

Springer

Editors
Sofia Brandão
Department of Radiology, CHSJ-EPE, Faculty of Medicine
University of Porto
Porto
Portugal

and

LAETA, INEGI, Faculty of Engineering
University of Porto
Porto
Portugal

Thuane Da Roza
Department of Biomechanics, College of Health and Sport Science (CEFID)
Santa Catarina State University (UDESC)
Florianópolis
Brazil

Isabel Ramos
Department of Radiology, CHSJ-EPE, Faculty of Medicine
University of Porto
Porto
Portugal

Teresa Mascarenhas
Department of Obstetrics and Gynecology, CHSJ-EPE, Faculty of Medicine
University of Porto
Porto
Portugal

ISSN 2212-9391 ISSN 2212-9413 (electronic)
Lecture Notes in Computational Vision and Biomechanics
ISBN 978-3-319-71573-5 ISBN 978-3-319-71574-2 (eBook)
https://doi.org/10.1007/978-3-319-71574-2

Library of Congress Control Number: 2017959158

Printed on acid-free paper

This Springer imprint is published by Springer Nature
The registered company is Springer International Publishing AG
The registered company address is: Gewerbestrasse 11, 6330 Cham, Switzerland

Foreword

This book relates to issues of Women's Health and Biomechanics. The book brings an interdisciplinary and multidisciplinary collaborative research to improve clinical outcomes in different aspects of women´s life. Experts from several fields contributed with a wide range of topics that affect women's health worldwide. The impact of the events occurring during the aging process in women's body and in its psychological and social impact is covered. This book included chapters related to physical activity or technologic platforms to fulfill well-being and to treat physical and psychological diseases, cancer-related mastectomy and its impact on body posture, breast reconstruction and implants, the biomechanics of pregnancy, the disorders of the pelvic floor (*postpartum*, due to age or among female athletes), sexual satisfaction and the quality of life in women with urinary incontinence, and its rehabilitation.

The chapters were elaborated considering the main body areas affected by the female life events, and, in this sense, this book gathers the major clinical and bioengineering perspectives from different professionals.

I hope the readers of this book will see the complex alterations which woman are subjected to during their life and will be able to use these concepts to understand the clinical and biomechanical practice and tools described here, and perhaps in a near future to apply them in subject-specific scenarios.

As one of the organizers of the related conference "International Conference on Clinical and BioEngineering for Women's Health (BioMedWomen)", that took place on June 20–23, 2015, in Porto, Portugal, which was the genesis for the present contribution, I would like to express my personal gratitude for the editors of this book and also to all authors for sharing their work and their knowledge in the context of the women's health and well-being.

Renato Natal Jorge
University of Porto, Porto Portugal
e-mail: rnatal@fe.up.pt

Contents

Editors and Contributors

About the Editors

Isabel Maria Ramos, MD, Ph.D. received her MD degree from the Faculty of Medicine of the University of Oporto. She completed a fellowship at Yale University in 1989, and in the same year she received her Ph.D. from the University of Porto. Since 1990, she works in the Centro Hospitalar de São João-EPE, in Porto, as Director of the Radiology Department, and since 2007, she accumulates with the autonomous unit of management. She was Chairman of the Quality Programme of the Centro Hospitalar de S. João-EPE for 5 years and Chairman of the Administration Board in 2003 and 2004.

She is also Full Professor at the Faculty of Medicine of the University of Porto since 2001, besides of being in the past Chairman of the Faculty of Medicine Scientific Board and of the Postgraduate Institute. She is also a member of the Portuguese Medical Academy.

In the context of Radiology Societies, she was already (Vice-) and President of the Portuguese Society of Radiology, and a member of the Board (and President) of the Portuguese College of Radiology. From 1996 to 2000, she was a member of the Board of the European Association of Radiology Education Committee. She is also a Portuguese representative at Union Européenne des Médecins-Spécialistes (UEMS—Radiology) since 1995.

She is (co)-author of more than 200 papers published in Portuguese and international journals, nine chapters in Radiology books, and performed more than 300 lectures.

In 2016, she was awarded the medal of merit of the Portuguese Order of Physicians.

Sofia Brandão, RT, Ph.D. is a radiographer at the Centro Hospitalar de São João-EPE. She received her B.Sc. and Graduation in Radiological Sciences by the School of Allied Health Sciences of the Polytechnic Institute, a Postgraduate and Ph.D. in Biomedical Engineering by the Faculty of Engineering of the University of Porto, and her M.Sc. in Medical Informatics by the Faculty of Medicine of the University of Porto. For the past 13 years, she is an Assistant Professor at the CESPU, CRL—Advanced Institute of Health Sciences, in Porto. Her teaching activities are developed in different B.Sc. and Postgraduate courses of Health Sciences.

She is a member of the EFRS—European Federation of Radiographer Societies, and has participated for several times as invited speaker and moderator in the annual European Congress of Radiology (ECR) of the European Society of Radiology (ESR), and also on the Annual Congress of the European Society for Magnetic Resonance in Medicine and Biology (ESMRMB).

The main research interest is the field of magnetic resonance (MR) imaging applications, namely in brain, breast, and pelvic MR, and also pelvic MRI for computational modeling. She has participated in different Ph.D. and Postdoctoral Projects in the context of MRI protocol optimization and acquisition.

Sofia Brandão has received a Training Scholarship at the University Hospital of Tübingen, which is an International Reference Center of Siemens Medical Systems, in 2003, and several awards in national and international conferences. She has over 20 peer-reviewed publications and has participated in a book chapter. Furthermore, she has performed more than 30 talks on technical aspects of clinical MRI, as well as on imaging and computational analysis of the pelvic floor.

Teresa Mascarenhas Saraiva, MD, Ph.D. received her MD and her Ph.D. from the University of Porto. Since 1992 she works in the Department of Obstetrics and Gynecology of the Centro Hospitalar de São João-EPE, in Porto, and since 2006 and she is Head of Department and Associate Professor of the Faculty of Medicine of the University of Porto, respectively.

In the context of Professional and Scientific Societies, she was already the European Representative, member of the Scientific Board, and President of the International Urogynecological Association (IUGA), and was also a member of the Board of the European Urogynecological Association (EUGA). She was also founder of the Portuguese Section of Urogynecology of the Portuguese Society of Gynecology.

Professor Mascarenhas has developed relevant research work and has received several international and national awards. She is (co)-author of more than 200 publications, including book chapter and scientific papers in national or international journals, and has more than 400 conference presentations. She has also been (co-)supervisor of several M.Sc., Ph.D, and Postdoctoral theses in her research field, and is a member of the Editorial Board of two major International Journals in her area of expertise.

Thuane Da Roza, PT, Ph.D. is Professor of Obstetrics and Gynecology at the Santa Catarina State University (UDESC) with appointment as Staff in the Biomechanics Laboratory of Health Sciences and Sports Center. She received her B.Sc. and Graduation in Physiotherapy Science by the College of Health and Sport Science of the UDESC, a M.Sc. and a Ph.D in Physical Activity and Health by the Faculty of Sport at the University of Porto. She is currently completing a Postdoctoral fellowship at the College of Health and Sport Science, Florianópolis, Brazil, in conjunction with the Institute of Science and Innovation in Mechanical and Industrial Engineering of Faculty of Engineering of the University of Porto, Portugal.

She has over 30 peer-reviewed publications, has participated in a book chapter, and performed over 50 lectures regarding pelvic floor function and biomechanics. The focus of her 10-years research career has been on 3D pelvic floor structure and function, through computational modeling, especially in female athletes. She was the first to demonstrate the effect of a comprehensive pelvic floor muscles training program on urinary incontinence symptoms in sports women, as well as the differences in pelvic floor computational simulations between continent and incontinent athletes.

Contributors

Henrique A. Almeida School of Technology and Management, Polytechnic Institute of Leiria, Leiria, Portugal

Sandra Amado CIPER – LBMF, Faculty of Human Kinetics, University of Lisbon, Lisbon, Portugal; CDRSP, Polytechnic Institute of Leiria, Leiria, Portugal

Ana Amaral ESTeS Coimbra, Polytechnic Institute of Coimbra, Coimbra, Portugal

Paulo Armada-da-Silva CIPER – LBMF, Faculty of Human Kinetics, University of Lisbon, Lisbon, Portugal

Rita M. T. Ascenso School of Technology and Management, Polytechnic Institute of Leiria, Leiria, Portugal; Center for Research in Informatics and Communications, Polytechnic Institute of Leiria, Leiria, Portugal

Tânia Bastos Center of Research, Education, Innovation and Intervention in Sport (CIFI2D), Faculty of Sport, University of Porto, Porto, Portugal; Research Center in Sports Sciences, Health Sciences and Human Development (CIDESD), ISMAI - Instituto Superior Da Maia, Porto, Portugal

Bary Berghmans Pelvic Care Center Maastricht, Maastricht University Medical Center, Maastricht, The Netherlands

Aroj Bhattarai Biomechanics Laboratory, Institute of Bioengineering, FH Aachen University of Applied Sciences, Jülich, Germany

Sofia Brandão Department of Radiology, Centro Hospitalar de São João—EPE, Faculty of Medicine, University of Porto, Porto, Portugal; LAETA, INEGI, Faculty of Engineering, University of Porto, Porto, Portugal

Kari Bø Department of Sports Medicine, Norwegian School of Sports Sciences, Oslo, Norway

Alice Carvalhais Department of Physiotherapy, CESPU—North Polytechnic Institute of Health, Paredes, Portugal

Luis Carrão CIPER – LBMF, Faculty of Human Kinetics, University of Lisbon, Lisbon, Portugal; ESSLEI, Polytechnic Institute of Leiria, Leiria, Portugal

Rui Corredeira Research Center in Physical Activity, Health and Leisure (CIAFEL), Faculty of Sport, University of Porto, Porto, Portugal

Raquel Costa Research Center in Physical Activity, Health and Leisure (CIAFEL), Faculty of Sport, University of Porto, Porto, Portugal

Margarida Espanha CIPER – LBMF, Faculty of Human Kinetics, University of Lisbon, Lisbon, Portugal

Paula Faria CDRSP, Polytechnic Institute of Leiria, Leiria, Portugal; ESTG, Polytechnic Institute of Leiria, Leiria, Portugal

Antonio Augusto Fernandes INEGI, Faculty of Engineering of University of Porto, Porto, Portugal

Lydia M. Ferreira Chairwoman of the Division of Plastic Surgery, Federal University of São Paulo, São Paulo, Brazil

Tomas Fortoul Plastic Surgery Unit, Centro Médico Docente La Trinidad, Caracas, Venezuela

Ralf Frotscher Biomechanics Laboratory, Institute of Bioengineering, FH Aachen University of Applied Sciences, Jülich, Germany; TWT GmbH Science and Innovation, Stuttgart, Germany

Wendy L. Gilleard School of Health and Human Sciences, Southern Cross University, Lismore, NSW, Australia

Eluana Gomes Research Center in Physical Activity, Health and Leisure (CIAFEL), Faculty of Sport, University of Porto, Porto, Portugal

M. Graça Pereira Department of Applied Psychology, School of Psychology, University of Minho, Braga, Portugal

Filipa João Fac. Motricidade Humana, CIPER, LBMF, Univ Lisboa, Lisbon, Portugal

Soraia Cristina Tonon da Luz Department of Physiotherapy, Santa Catarina State University, Florianópolis, Brazil

Maria da Luz Barroso Department of Plastic Surgery, Centro Hospitalar de Gaia-Espinho—EPE, Vila Nova de Gaia, Portugal

Thalissa Maniaes Department of Health Science, Post-Graduate and Physical Therapy, University Santo Amaro, São Paulo, Brazil

Pedro Martins INEGI, Faculty of Engineering of University of Porto, Porto, Portugal

Teresa Mascarenhas Department of Gynecology and Obstetrics, Centro Hospitalar de São João—EPE, Faculty of Medicine, University of Porto, Porto, Portugal

Fabio X. Nahas Division of Plastic Surgery, Federal University of São Paulo, São Paulo, Brazil

Patrícia Mota Fac. Motricidade Humana, CIPER, LBMF, Univ Lisboa, Lisbon, Portugal; ESTeSL, Escola Superior de Tecnologia da Saúde de Lisboa, Instituto Politécnico de Lisboa, Lisbon, Portugal

Adriana Naomi Hamamoto Cancer Institute Hospital, Faculty of Medicine, University of São Paulo, São Paulo, Brazil

Renato Natal Jorge LAETA, INEGI, Faculty of Engineering, University of Porto, Porto, Portugal

Luísa Nogueira Department of Radiology, School of Health, Polytechnic of Porto (ESS-P.Porto), Porto, Portugal

Rita G. Nunes LARSyS, Department of Bioengineering, Institute for Systems and Robotics, Instituto Superior Técnico, Universidade de Lisboa, Lisboa, Portugal; Faculty of Sciences, Institute of Biophysics and Biomedical Engineering (IBEB), University of Lisbon, Lisbon, Portugal

Dulce Oliveira LAETA, INEGI, Faculty of Engineering, University of Porto, Porto, Portugal

Eunice Oliveira School of Technology and Management, Polytechnic Institute of Leiria, Leiria, Portugal; R&D Unit INESC Coimbra, Coimbra, Portugal

Marco Parente LAETA, INEGI, Faculty of Engineering, University of Porto, Porto, Portugal

Augusto Gil Pascoal Fac. Motricidade Humana, CIPER, LBMF, Univ Lisboa, Lisbon, Portugal

Susana Pedras Department of Applied Psychology, School of Psychology, University of Minho, Braga, Portugal

Michel Probst Research Group for Adapted Physical Activity and Psychomotor Rehabilitation, Faculty of Kinesiology and Rehabilitation Sciences, KU Leuven, Leuven, Belgium

Nilza Ramião INEGI, Faculty of Engineering of University of Porto, Porto, Portugal

Isabel Ramos Department of Radiology, Centro Hospitalar de São João—EPE, Faculty of Medicine, University of Porto, Porto, Portugal

Heather Lynn Rogers Biocruces Health Research Institute, Barakaldo, Spain; IKERBASQUE Basque Foundation for Science, Bilbao, Spain

Thuane Da Roza Department of Physiotherapy, Santa Catarina State University, Florianópolis, Brazil

Ana Paula Ribeiro Department of Physical Therapy, Faculty of Medicine, University of São Paulo, São Paulo, Brazil; Department of Health Science, Post-Graduate and Physical Therapy, University Santo Amaro, São Paulo, Brazil

Cinara Sacomori School of Kinesiology, Universidad Bernardo O'Higgins, Santiago de Chile, Chile

Fátima Sancho The Alcoitão School of Health, Alcabideche, Portugal

Diana Costa Santos Department of Plastic Surgery, Centro Hospitalar de Gaia-Espinho—EPE, Vila Nova de Gaia, Portugal

Rute Santos CIPER – LBMF, Faculty of Human Kinetics, University of Lisbon, Lisbon, Portugal; ESTeS Coimbra, Polytechnic Institute of Coimbra, Coimbra, Portugal

Maura Seleme abafi-HOLLAND, Maastricht, The Netherlands; Faculdade Inspirar, Curitiba, Brazil

Cláudia Senra Department of Applied Psychology, School of Psychology, University of Minho, Braga, Portugal

Gesilani Júlia da Silva Honório Department of Physiotherapy, Santa Catarina State University, Florianópolis, Brazil

Elisabete Silva LAETA, INEGI, Faculty of Engineering, University of Porto, Porto, Portugal

Patrícia Silva ESTeS Lisboa, Polytechnic Institute of Lisbon, Lisbon, Portugal

Manfred Staat Biomechanics Laboratory, Institute of Bioengineering, FH Aachen University of Applied Sciences, Jülich, Germany

Cristiana Vaz Fac. Motricidade Humana, CIPER, LBMF, Univ Lisboa, Lisbon, Portugal

António Veloso CIPER – LBMF, Faculty of Human Kinetics, University of Lisbon, Lisbon, Portugal

Fabiana Zanata Division of Plastic Surgery, Federal University of São Paulo, São Paulo, Brazil

Introduction

The definition of health, as described by the World Health Organization (WHO), includes physical, mental, and social well-being, and not just the absence of disease or infirmity. In this context, women experience unique health issues and concerns. The aging process of youth to old age, pregnancy, the bearing of children, as well as the hormonal changes in menopause, can lead to several pathologies such as breast cancer or pelvic floor dysfunction, and affect women worldwide.

Health care includes evaluating the psychological and social impacts that most clinical conditions imply. On the other hand, mental illness itself or chronic diseases can significantly reduce quality of life. Biochemical changes are well known and described in the diagnosis and treatment of physical distress due to aging or debilitating illness. Furthermore, psychological suffering—which can often cause hormonal alterations and even the alteration of neurotransmitters—may also constitute a health problem. Clinical and lifestyle changes such as physical activity or quality-of-life self-management can be supporting to the more classical medication prescription approach in such cases.

In the context of women's health, anamnesis (or medical history), clinical evaluation, and the choice of the most suited treatment options, whether conservative or surgical, is indeed the most common approach. In this context, to better understand the phenomena leading to the pathologic condition, diagnostic tools such as imaging techniques, imaging, or biomechanical-based computer analysis are often applied. Furthermore, medical devices or prostheses are currently available to be used during treatment, which means bringing different branches of Engineering to the Clinical environment in female health.

Having this in mind, our purpose is to present some of the aspects of women's health, some of which are clearly better understood and treated by bridging knowledge from Medicine and Engineering.

Section 1, entitled "Women's Health", emphasizes the psychological aspects of female health, such as those regarding the application of information technologies to monitor and assist women with chronic disease such as fibromyalgia (Chapter "Improving Women's Health via the Biopsychosocial Model: Fibromyalgia as a Case Study to Explore Opportunities for Engineering Applications") or the impact of mental illness and the positive effect of physical activity programs in states of (post-natal) depression or eating disorders (Chapter "Physical Activity and Women's Mental Health"). Pain and reduction of quality of life are also often described in the elderly (female) population. It relates to deterioration of the musculoskeletal system leading to osteoarthritis in large joints such as hips, shoulders, and knees, mostly related to changes in the endocrine system, and also associated with the postmenopausal period. By investigating muscle architecture and properties by means of ultrasound (Chapter "The Role of Ultrasound Imaging of Musculotendinous Structures in the Elderly Population"), loss of muscle strength and functional decline can be assessed early, and the most suited geriatric rehabilitation programs can be developed.

Section 2, entitled "Bridging Clinical and Biomechanical Aspects of Breast Surgery", addresses the subjects of breast cancer diagnosis (Chapter "Diffusion-Weighted Breast Imaging: Beyond Morphology"), rehabilitation (Chapters "The Effects of Mastectomy and Breast Reconstruction on Body Posture and Biomechanical Aspects and The Role of Physiotherapy in Female Breast Cancer"), and surgical implants (Chapter "Breast Implants: Far Beyond Just Aesthetic Surgery"). Breast cancer is the most common cancer in women worldwide, with nearly 1.7 million new cases diagnosed in 2012. In Europe, Belgium had the highest rate of breast cancer, but also the highest proportion of 5-year survivors, followed by Denmark and France. An early and accurate diagnosis is a key factor, and Magnetic Resonance Imaging has proved to acquire information regarding microstructural complexity of both normal breast tissue and lesions non-invasively and with great accuracy (Chapter "Diffusion-Weighted Breast Imaging: Beyond Morphology"). After surgery, rehabilitation is usually performed, aiming for the promotion of health, improving wellness and quality of life in daily life activities (Chapter "The Role of Physiotherapy in Female Breast Cancer"). An important aspect to consider is the fact that there are adaptations of the musculoskeletal system related to mastectomy and breast implants. These biomechanical posture changes may provide relevant information for prescribing physical exercise and rehabilitation programs (Chapter "The Effects of Mastectomy and Breast Reconstruction on Body Posture and Biomechanical Aspects").

Breast implants are widely used in cancer patients after radical mastectomy or for breast augmentation purposes. Some adverse effects have been described, which is a major concern for patients, plastic surgeons, manufacturers, and regulatory agencies. There has to be certainty on the safety and compliance of the implants used. Detailed visual analysis and product testing provide important details, for example, on the mechanical properties of the materials inside the implant shell, and loads to which they may be subjected (Chapter "Breast Implants: Far Beyond Just Aesthetic Surgery").

Section 3, entitled "The Biomechanics of the Reproductive Period", focuses common conditions associated with pregnancy and childbirth. During this period, body composition and posture change significantly, and one can define an anthropometric profile specific of this stage of a woman's life. Chapter "Anthropometrics and Ergonomics in Pregnant Women" discusses practicing good ergonomics at home and in the workplace, which will provide comfort and improve women's health throughout this period. During pregnancy, the abdomen becomes larger, the curvature of the lumbar spine increases, and the gait changes. The changes seen in step width are thought to reflect mechanical rather than functional adaptation to increase stability, which is discussed in Chapter "Increased Step Width during Walking as Pregnancy Progresses: Functional or Mechanical Adaptation?".

Another important issue in this period is the well-known *diastasis recti abdominis*, which develops during pregnancy and may be present during the first weeks after childbirth. The prevalence and risk factors for development of this condition are still under study. In Chapter "Diastasis Recti during Pregnancy and Postpartum ", the authors discuss the application of early diagnosis by means of ultrasound as an easy way to plan an effective exercise program to reduce the *inter rectus distance*. Another frequent condition in *postpartum* is pelvic floor dysfunction, namely urinary incontinence. This may result from changes in the connective tissue due to hormonal changes or direct muscle damage during the traumatic event of vaginal childbirth. For that purpose, computational modeling has been used to study the biomechanics of the pelvic floor. Several subject-specific factors have to be taken into account. Structural congruence between shape and position of the fetal head at delivery with the women's pelvic girdle and pelvic floor muscles' shape is important, but also the soft tissue properties, and damage analysis during the different stages of vaginal delivery. These inputs are significant to understand the pathophysiology of the resultant functional damage (Chapter "Biomechanical Analysis of the Damage in the Pelvic Floor Muscles During Childbirth").

Pelvic floor dysfunction is indeed one of the most prevalent conditions during women's life. Hence, we gave it a great deal of attention and detail in the present book. In this particular field, much experimental, clinical, and mechanical research has been carried out in different branches of the subject. From *postmortem* tissue experimental tests, computational simulation, to elaborating and testing the most suited surgical materials, much effort has been made to understand the underlying conditions that promote developing pelvic floor dysfunction. Sections 4 and 5 focus on different aspects of studying pelvic floor dysfunction.

Section 4 entitled "Clinical Approach on the Female Pelvic Floor" was thought to emphasize clinical evaluation of the pelvic floor muscles and the impact of urinary incontinence in quality of life. Not only elderly or parous women suffer from urinary incontinence but also female athletes are at risk. Exercise-induced urinary incontinence is not yet fully understood, but it is thought to result not only from sports practice itself but also from individual factors. Chapter "Pelvic Floor in Female Athletes: From Function to Dysfunction" reviews these issues and sheds light on the impact of urinary incontinence on the quality of life of young athletes, and on the relevance of developing effective preventive physiotherapy. When evaluating the function of the pelvic floor muscles, measuring vaginal squeeze pressure is a means for assessing muscle strength. Chapter "Towards the Development of a Vaginal Finger-cot Device for Measuring Pelvic Floor Muscles Strength" presents a customized adjustable vaginal finger-cot device that can be used to provide biofeedback and motivate women during the rehabilitation protocol. In this same context, Chapter "Physiotherapeutic Diagnostic Process for Female Urinary Incontinence" describes the rationale and the steps of a clinical pelvic diagnostic process for female urinary incontinence toward an adequate pelvic physiotherapy evaluation and treatment.

The impact of urinary incontinence on women's quality of life has to be thought. The psychological impact in her well-being, coping mechanism and sexual life are important factors to account to when developing the rehabilitation process itself, as well as health promotion strategies, as explained in the Chapter "Psychological Morbidity, Sexual Satisfaction, Coping, and Quality of Life in Women with Urinary Incontinence in Rehabilitation Treatment".

The last part of the book relates to the "Biomechanical Analysis of the Female Pelvic Floor" (Section 5). Until now, significant advances have been made to characterize biomechanical properties of pelvic floor structures to identify possible mechanisms that contribute to developing pelvic floor dysfunctions. Several studies have been performed to improve and to identify the most suited constitutive models for the pelvic floor muscles. While Chapter "Computational Analysis of Pelvic Floor Dysfunction" presents an integrated discussion of the constitutive model based on the tissue constituents, fiber, and ground substance, Chapter "Searching for the Tissue Mechanical Properties in Pelvic Floor Dysfunction by Computational Modeling" presents a method that allows estimating the most suited material constants for the pelvic floor muscles for a subject-specific using input information for the computational simulations acquired non-invasively: the inverse finite element analysis. Accordingly, we expect that this book section can clarify and help future researches to choose the best constitutive model and understand this complex condition.

We would like to recognize the effort and time spent by the authors to these manuscripts, as well as acknowledge serenity for the comments of the editors and for the valuable work performed. We also would like to thank the editorial team of Springer, who carefully worked hard on every detail. Finally, we hope our book will help all readers to understand or improve their skills to assist women's health.

Part I
Women's Health

Improving Women's Health via the Biopsychosocial Model: Fibromyalgia as a Case Study to Explore Opportunities for Engineering Applications

Heather Lynn Rogers

Abstract The biopsychosocial model of health provides a framework to assess and/or treat various medical disorders and is the most heuristic approach to managing chronic pain. Fibromyalgia (FM) is a chronic pain disorder primarily affecting women characterized by widespread musculoskeletal pain, abnormal pain processing, sleep disturbance, fatigue, and often cognitive difficulties and psychological distress. Evidence-based management guidelines in different countries recommend biopsychosocial, lifestyle-oriented intervention to include exercise, cognitive-behavioral therapy, and multicomponent intervention. State-of-the-art evaluation and treatment approaches in FM illustrate the application of the biopsychosocial model to improve women's health. Engineering applications are beginning to be developed, that, within the context of this model, have the potential to further advance management of the disorder and improve quality of life for those (primarily women) who suffer from it. Users of existing online multicomponent treatment module platforms experience improved pain and physical functioning. Those who use FM symptom tracking systems report improvements in a number of debilitating core FM symptoms beyond pain and physical function. Areas of engineering application showing particular promise include advances in gaming using commercially available motion-controlled video games, mobile activity and symptom data collection with or without feedback via smartphone devices, and integration of these technologies with clinical oversight.

Keywords Biopsychosocial model · Evidence-based guidelines
Fibromyalgia · Engineering/IT applications · Quality of life

H. L. Rogers (✉)
Biocruces Health Research Institute, Barakaldo, Spain
e-mail: rogersheatherl@gmail.com

H. L. Rogers
IKERBASQUE Basque Foundation for Science, Bilbao, Spain

S. Brandão et al. (eds.), *Women's Health and Biomechanics*,
Lecture Notes in Computational Vision and Biomechanics 29,
https://doi.org/10.1007/978-3-319-71574-2_1

Highlights

- Fibromyalgia (FM) is a chronic pain disorder primarily affecting women and evidence-based guidelines recommend biopsychosocial, lifestyle-oriented intervention for management of the disorder.
- Existing engineering applications have been developed that have demonstrated the potential to further advance self-management of the disorder and improve quality of life for those (primarily women) who suffer from it.
- Specifically, users of existing online multicomponent treatment module platforms experience improved pain and physical functioning and those who use FM symptom tracking systems report improvements in a number of debilitating core FM symptoms beyond pain and physical function.
- Areas of engineering application showing particular promise include advances in gaming using commercially available motion-controlled video games, mobile activity and symptom data collection with or without feedback via smartphone devices, and integration of these technologies with clinical oversight.

1 Introduction

Since 1948, the World Health Organization (WHO) has defined health as "a state of complete physical, mental, and social well-being and not merely the absence of disease or infirmity" [1]. The biopsychosocial model [2] provides an appropriate framework to promote general health/well-being. This biopsychosocial model is now a widely accepted approach to the assessment and/or treatment of many disorders in various medical subspecialties. In particular, in the case of chronic pain conditions, the biopsychosocial model is now viewed as the most heuristic approach to management [3].

This chapter has five objectives:

(1) To describe the biopsychosocial model
(2) To place the biopsychosocial model in the context of the history of Western Medicine
(3) To apply the biopsychosocial model to the evaluation and treatment Fibromyalgia (FM), a chronic pain disorder primarily affecting women,
(4) To review examples and evidence-based benefits of engineering/information technology (IT) applications in the management of FM, and
(5) To highlight new developments in engineering/IT to assist those with FM.

2 The Biopsychosocial Model

In 1977, George Engel published a seminal paper, "The Need for a New Medical Model: A Challenge for Biomedicine" [2], in which he critiques the biomedical model prevalent at the time. He describes the requirements of a new medical model to account for disease as a human experience, and proposes a biopsychosocial model as a "blueprint for research, a framework for teaching, and a design for action in the real world of health care" [2, p. 135]. Engel proposes a new dynamic and interactional medical paradigm that broadened the scope of the clinician's focus in diagnosing and treating disease. As a consequence, he expands the domain of medical knowledge to address the needs of each individual patient.

Borrell-Carrio, Suchman, and Epstein [4] summarize the key tenants of Engle's Biopsychosocial Model:

(1) Illness results from the interaction of several diverse causes, including those at the molecular, individual, and social levels. A biochemical alteration does not necessarily translate directly into an illness, and psychological alterations may, under certain circumstances, manifest as illnesses or forms of suffering that constitute health problems, including, at times, biochemical correlates.
(2) Psychosocial factors influence the susceptibility, severity, and course of illness, as well as the effect of treatment.
(3) Determining the biological cause does not necessarily infer meaning to the patient. Furthermore, the success of the most biological treatments is influenced by psychosocial factors, e.g., the so-called placebo effect.
(4) The patient-physician relationship influences medical outcomes, at minimum as a result of its influence on adherence to the selected treatment.
(5) Adopting a sick role is not necessarily associated with a biological cause

Engel identified weaknesses in the biomedical model and defined a multidisciplinary approach to treating illness known as the "Biopsychosocial Model." This model proposes that a combination of factors play a significant role in human functioning in the context of disease—not only the biological factors in the biomedical model of illness, but psychological (which includes thoughts, emotions, and behaviors) and social (socioeconomical, socio-environmental, and cultural) factors as well [5]. In summary, the main distinction between the biomedical model and the biopsychosocial model is the full integration of "psychosocial" factors. These psychosocial factors are the same health determinants that contribute to the development and progression of many noncommunicable, or lifestyle, diseases involving all of the body's systems (e.g., coronary heart disease, musculoskeletal disorders, Type 2 diabetes, cancer, asthma, Alzheimer's disease). In fact, studies have shown that when physicians address the patient's lifestyle and environmental context via use of the biopsychosocial model, they are more likely to provide more psychosocial advice/interventions and prescribe fewer medications, with an increase in patient-reported satisfaction [6].

3 A Brief History of Western Medicine

An examination of the history of Western medicine's conceptualizations regarding the mind-body relationship in health and disease elucidates how the biopsychosocial model integrates factors that have been considered to be influential at various points in time. It is possible to argue that the biopsychosocial model brings Western medicine full circle to where it began with the ancient Greeks. In brief, in the early days of medicine, illness was almost always explained in spiritual terms. Hippocrates, often recognized as the father of Western medicine, proposed a new paradigm in which natural, not supernatural, explanations of illness were sought. Because dissection of human cadavers was forbidden on religious grounds, Hippocrates and Aristotle relied on logic and philosophy to explain disease. Centuries later, Galen, a Roman anatomist, studied pigs and integrated the role of personality in illness. Hippocratic-Galenic medicine hypothesized a "Holism." Under this view, there was a synergistic and individual relationship between a person's body, mind, personality, and the outside world. These factors are postulated as integrated and inseparable [7].

In the Middle Ages, the Catholic Church became a powerful political and social force. The "mind-body issue" was deemed religious and investigations into human anatomy were not allowed. Later in the 1500s, physicians were allowed to dissect executed criminals. Hence, structures that could only be imagined prior could be now manipulated to reveal clues about their function. By the mid-1600s, physicians began to view human physiology as the mechanized interaction of organs. Growth in medical technology played a key role in reductionist perspective in Western Medicine. For instance, the microscope revolutionized biology and was used as a tool to study not just simple organisms but also the disease process. The etiology of disease could be tied to distinct anatomical locations [7]. At the same time, during the Renaissance, Descartes popularized the belief of "Dualism"—a distinction between mind and body. Descartes' dualism separated the mind (soul), which was left to the Church, from the body, which was now available for physicians and scientists to study in this reductionist and mechanical manner.

In the eighteenth century, with the Industrial Revolution, it was possible to observe a shift from a purely mechanistic view of health to a more integrative view of health within the context of one's environment. Then, the nineteenth–twentieth century was characterized by a shift back to reductionism in order to understand illnesses at the time with the largest impact on public health (e.g., infectious diseases) and treat their mechanisms of action. In the twenty-first century, chronic illnesses are now the primary causes of mortality. As espoused by the biopsychosocial model, the mind-body interaction approach to the study and treatment of chronic illnesses is an integration of the holistic approach to early medicine with modern reductionist understanding of disease processes.

4 Fibromyalgia Evaluation and Treatment—Applying the Biopsychosocial Model in the Twenty-First Century

Fibromyalgia (FM), also known as Fibromyalgia Syndrome (FMS), is a chronic pain disorder that has a debilitating impact on women's health. FM is a condition characterized by widespread musculoskeletal pain that is accompanied by abnormal pain processing, sleep disturbance, fatigue, and often cognitive difficulties and psychological distress. Its etiology is unknown [8]. The disorder typically affects women 40–55 years of age, has a female-to-male ratio of 9-to-1, and can affect up to 5% of the population [9]. Considerable quantitative (e.g., Gonzalez et al. [10]) and qualitative (e.g., Arnold et al. [11]) research has documented the important impact that FM has on quality of life. Negative impact on social function includes social isolation and interference with family relationships, friendships, activities of daily living, leisure activities, and physical activity. Negative professional impact, such as the loss of career or inability to advance in careers or education, has also been reported.

The need for biopsychosocial assessment of FM has become even more apparent with the American College of Rheumatology (ACR) 2010 revised criteria for diagnosing FM [12] which are focused on patient perception of pain in 19 specific locations (known as the Widespread Pain Index; WPI) and six self-administered symptoms questionnaires to assess sleep difficulty, fatigue, headache, depression, abdominal pain, and poor cognitive status (known as the Symptom Severity Score; SSS).

As a multidimensional disorder, it is not surprising that multicomponent interventions involving a biopsychosocial perspective are the recommended treatment for FM. Evidence-based management guidelines have been developed by various associations in different countries, for instance, in Canada in 2012, by the Association of the Scientific Medical Societies (AMWF) in Germany in 2008 and the American Pain Society in 2005. Most recently, in 2016, the original 2007 European League Against Rheumatism (EULAR) recommendations for managing fibromyalgia were updated. Exercise was given a "strong" therapy-based recommendation based on meta-analyses. Although other therapies had "weak" evidence according to meta-analyses, expert opinion suggests a staged, individualized approach to FM treatment. Intervention was to begin with patient education and a focus on non-pharmacological therapies. If the individual does not respond to this management of FM, three additional therapies in the following order were recommended: (1) Psychological therapies to improve mood disorders and unhelpful coping strategies, (2) Pharmacotherapy to manage of severe pain or sleep disturbance, and/or (3) A multimodal rehabilitation program to address severe disability [13]. In general, guidelines from the United States, Europe, and Canada coincide in recommending aerobic exercise, cognitive-behavioral therapy, and multicomponent intervention as first-line therapies. There is also some agreement among experts regarding the need for a stepwise approach to the management of FM with an emphasis on longer term self-management of this incurable condition that provokes considerable suffering.

5 The Role of Engineering Applications in the Effective Management of Fibromyalgia

The “lifestyle-oriented” interventions recommended for the treatment of FM by clinical practice guidelines in various countries include patient education, aerobic or other physical exercise, cognitive-behavioral therapy (CBT), and multicomponent interventions. These approaches to FM management can be delivered in various forms—for instance, the individual with FM could regularly go to a medical or behavioral healthcare setting and receive instruction from trained professionals and/or he/she could use a self-help type patient-oriented treatment manual that is implemented at home with less frequent check-ins with healthcare professionals. Recently, remote-access technologies are being used more commonly to effectively deliver these types of interventions at the individual’s home and via the Internet or mobile technologies [14]. Advancements in IT allow for ease of access to lifestyle-oriented interventions that provide much more than basic information and guidelines (as might be found in a treatment manual). Heron and Symth [14] emphasize that current IT lifestyle-oriented intervention programs can offer interactive and adaptive responses for individual patients, based on real-time data transfer of information about the individual patient. Such intervention may be more cost-effective than face-to-face individual or group therapy sessions and can be realized in the individual’s own home at his/her convenience. Additionally, individuals can capitalize on online supportive interaction elements allowing them to connect with others suffering from their disorder. This feature can help address the loneliness, invalidation, and lack of social support that contribute to distress in people with FM [15].

Daraz et al. [16] conducted a review of the first 25 websites identified using a Google search of the term “fibromyalgia” found that approximately one-third were “very good,” one-third were “good”, and one-third were of “marginal” quality. Only 16% of websites met the recommended literacy level grade of 6–8 (range 7–15), with higher quality websites receiving lower readability ratings. The authors concluded that, although educational information is very important for individuals suffering from FM, existing online resources evaluated at the time were unlikely to provide sufficiently accurate information that could be understood by a majority of the general public.

Fortunately, a number of evidence-based biopsychosocial engineering/IT interventions to assist individuals with FM have been developed and studies of their efficacy seem to produce positive results. Friedberg, Williams, and Collinge [17] conducted a review of online interventions in FM highlighting these studies and IT contributions to improving quality of life in individuals with FM. Specifically, in a randomized controlled trial, a module-based Internet exercise and behavioral self-management program called Web-Enhanced Behavioral Self-Management (WEB-SM) program of clinician lectures, readings, homework, and self-monitoring forms was found to significantly improve pain and physical functioning, along with overall global improvement from baseline to 6-month follow-up in the experimental group using the website [18]. The "Living Well with Fibromyalgia" intervention website contained (1) educational lectures providing background knowledge about FM as a disease state; (2) education, behavioral, and cognitive skills designed to help with symptom management; and (3) behavioral and cognitive skills designed to facilitate adaptive lifestyle changes for managing FM. Exercise and relaxation techniques (including use of the available audio recordings for relaxation) were reported to be the most commonly used skills by those with FM. Another randomized trial of a 6-week online, Internet-based CBT program called MoodGYM had lower Fibromyalgia Impact Questionnaire (FIQ) scores and tender point scores at 6- and 12-week follow-ups compared to those receiving usual care. Within the MoodGYM group over time, there was a significant improvement in tender point scores from baseline to 6-weeks, which was maintained at 12-weeks, but no significant difference in the FIQ scores, suggesting that the program might help reduce pain in those suffering from FM [19].

As demonstrated above, one engineering/IT application in FM intervention is to provide web-based delivery of lifestyle-oriented intervention programs that have been proven to be effective when delivered by other means. Another engineering/IT application to potentially improve outcomes in individuals suffering from FM is to provide electronic means of logging information that, when fed back to the individual, can provide insight into relationships between stress, behaviors, and symptom levels that the individual can self-manage. Collinge, Soltysik, and Yarnold [20] developed and tested such a system, called "SMARTLog" (SMART, Self-Monitoring and Review Tool), for people with FM. Individuals with FM used an online instrument to record daily data on their activities, health-related behaviors, stressors, and symptom levels. As users used the system about 5 minutes per day for approximately 20 days, a personal database was built, from which cumulative data could be analyzed. Personal behavioral predictors of statistically significant changes in patients' symptom levels could be determined and appropriate recommendations inferred. For instance, the program might indicate: "When you eat dinner before 7:13 pm, your pain levels are lower the next day." The pilot feasibility study showed high utilization, satisfaction, and compliance during the testing period, with higher utilization predictive of lower anxiety and improved physical functioning and self-efficacy [20].

In a Phase II study of SMARTlog, Collinge, Yarnold, and Soltysik [21] found that likelihood of benefit was a function of the frequency of use of the program. Specifically, moderate users of the program (defined as 3 times weekly during 3 months) had an increased likelihood of clinically significant improvements in pain, memory, gastrointestinal problems, depression, fatigue, and concentration. Heavy users (defined as 4.5 times weekly during 5 months) had increased likelihood of clinically significant improvements in the above areas, plus improvement in stiffness and sleep difficulties. It is of note that these outcomes were based on simple utilization of the SMARTLog system, regardless of the SMART Profile statements targeting specific behavior–symptom relationships. When the user's data produced specific SMART Profile "target" statements, the impact of the intervention on specific symptoms was found to be greater. The SMARTLog program is planned to be made available at the URL: www.awarehealth.org.

6 New Engineering/IT Applications Recently Developed to Assist Those with Fibromyalgia

As described in the prior section, there have been numerous high-quality studies with large numbers of participants (in some cases over 800) providing an evidence-base for the moderate benefit of engineering/IT applications in the management of FM. These existing approaches are capable of delivering substantive information and guidance, even personalized feedback systems, and interactive opportunities that begin to effectively address and improve some of the difficulties that individuals with FM face. They can be used with or without a professional healthcare provider's oversight. However, much more engineering/IT development could be undertaken, and preliminary pilot studies suggest added value in the management of FM in the areas of gaming, smartphone data collection and monitoring by healthcare professionals, integration of electronic health records to improve patient care, and new wearable monitoring technologies. Each of these will briefly be described in the following subsections below.

6.1 Gaming

A study of 15 women diagnosed with FM by Mortensen et al. [22] suggests that Motion-Controlled Video Games (MCVGs) may be an acceptable intervention for people with chronic pain. In the study, the participants played five sessions of each of three commercially available MCVGs: Nintendo Wii (Wii), PlayStation 3 Move (PS3 Move), and Microsoft Xbox Kinect (Xbox Kinect). Participants reported being distracted from their pain symptoms. Xbox Kinect was reported as the best console for exercise. In general, they enjoyed the slow pace and familiarity of Wii, while some considered PS3 Move to be too fast-paced. In this small sample, there was no evidence of improvement in symptom severity or performance of Activities of Daily Living (ADL). Additional development of low impact exercise modules using Xbox Kinect tested in larger samples may show clinical benefit, and this is an area ripe with opportunity to make an important impact on women's health [22].

6.2 Smartphone Data Collection and Monitoring

Electronic diaries completed using a smartphone may provide means of ecological momentary assessment (EMA) that allows for repeated assessment of the same symptoms in real-time, eliminating the retrospective bias associated with paper diary completion. In a crossover study design by Garcia-Palacios et al. [23], 47 individuals with FM from Spain were randomly assigned to either (1) paper diary—smartphone diary or (2) smartphone diary—paper diary, using each assessment method for 1 week. The smartphone diary ratings were found to be more accurate and complete than the paper diary ratings and the method was well-accepted even though many of the participants had low educational levels and low familiarity with technology. Such real-time data collection technology has the potential for even more positive impact if it is integrated with clinical services. In a nursing study by Vanderboom et al. [24], 20 individuals with FM used a mobile monitoring device for 1 week and nurses responded to patient e-mailed symptom reports on a daily basis. Notably, participants wanted to use a smartphone to monitor their health and to communicate with health care providers. They used the study mobile device an average of 5.2 days out of the 7-day study period. 80% reported that monitoring symptoms using the device was easy to do and 65% indicated that using the device helped them to promptly address their symptoms. Further development of such technology integrated with clinical care monitoring could empower self-management of FM and decrease health care utilization and costs.

6.3 Integration of Electronic Health Records

Given the biopsychosocial nature of FM, effective management tends to require multidisciplinary intervention. An individualized FM management framework based on patient education and goal setting has become the *status quo*. Wells et al. [25] suggested that integration of patient electronic health records across providers could be incorporated at every stage of patient care, from initial presentation to diagnosis, through to making treatment decisions and maintaining ongoing patient management. Such integration is expected to improve patient care, create greater efficiencies in primary care provider management, and reduce healthcare system costs. Engineering/IT approaches can help to optimize intervention and help to drive best-practice care for individuals with FM while improving patient outcomes.

6.4 Wearable Monitoring Technologies

Possibly the area with the most potential for engineering/IT applications relates to the capture of physiological data relevant to specific disorders like FM.

For instance, sleep disturbance is one of the cardinal symptoms of FM, yet sleep diagnosis is typically done in a laboratory where patients are asked to stay overnight so that various types of physiological signals can be recorded. In-home evaluation is preferred, as it is more conducive to natural sleep and would, therefore, allow a more accurate capture of disease symptoms. Kayyali et al. [26] developed and tested PSG@Home, a 14-channel wearable wireless monitor and a cell phone-based Gateway to transfer data, including video, in real-time from the patient's home to a remote laboratory where data can be monitored and scored by a sleep specialist in real-time. The technology was piloted 10 individuals with FM whose constant pain has previously made them reluctant to travel to a sleep lab. All 10 studies were successful and generated high-fidelity recordings. Although one participant experienced intermittent real-time data transmission due to sparse cellular coverage, the data were recovered from a backup memory housed inside the patient monitor. No disconnections in sensor lead wires occurred. These preliminary results are encouraging, but further research is needed with this new technology to generate conclusions about sleep quality.

In a second example, technological advancements in the measurement of cardiac coherence real-time could help individuals with FM, and other, learn breathing exercises or other relaxation techniques that could help them control stress and autonomic nervous system activation that potentiates chronic pain. De Jockheere et al. [27] developed a smartphone-based solution that allows people to process efficient cardiac coherence biofeedback exercises. The authors used sensorless technology, and specifically photo-plethysmographic imaging through the smartphone camera, and then apply a heart-rate variability algorithm to the signals to facilitate reporting of cardiac coherence during biofeedback exercises. One benefit of this type of biofeedback available via portable, mobile technology is that the individuals learn techniques and stress management in the context of real-life situations, not in the lab or at home, and can visually see the changes in their physiology to verify the subjective improvements in well-being they experience.

7 Conclusion

Users of existing online platforms that deliver evidenced-based multicomponent treatment modules for FM experience improved pain and physical functioning and those who use FM symptom tracking systems like SMARTlog report improvements in a number of core FM symptoms beyond pain and physical function that are debilitating and have an important impact on quality of life. These benefits, demonstrated in studies with rigorous methodologies and large sample sizes, are quite encouraging. Furthermore, there are numerous new technologies being developed that could have even greater impact on those suffering from FM, thus leading to overall enhancement of women's health in general. Areas of engineering/ IT applications showing particular promise include advances in gaming using commercially available motion-controlled video games, mobile activity and symptom data collection with or without feedback via mobile smartphone devices, and integration of these technologies with clinical oversight and/or monitoring. Existing sensors in smartphones or fitness accessories, like location indicators, accelerometers, and step-counters, could be leveraged to offer automatic data collection for individuals with FM and then combined the information with self-reported symptom reports in real-time. Furthermore, new sensor-based technologies can be developed (e.g., to measure autonomic nervous system activity, sleep, etc.) to noninvasively log physiological changes and determine what activities are associated with positive changes in these indicators. Few technological platforms or interventions have taken full advantage of the social connections that can arise from online groups and collaboration, which is an additional value-added attribute that deserves research attention, especially in this population of women.

Engineers and those working in IT are key stakeholders who must be engaged in facilitating these positive changes in women's health, and this book represents one step in the process. The technological developments described in this chapter, when guided by a user-centered approach that takes into account the needs of individuals with FM regardless of their technological ability, have the potential to improve symptom management of women with chronic pain conditions like FM and also, as general lifestyle-oriented interventions, empower women to take control of their own health and enhance their own well-being.

References

1. Constitution of the World Health Organization (2006) World Health Organization: Basic documents. 46th edn. Geneva: World Health Organization
2. Engel GL (1977) The need for a new medical model: a challenge for biomedicine. Science 196(4286):126–136
3. Gatchel RJ (2013) The biopsychosocial model of chronic pain. In: Liegey Dougal A, Gatchell RJ (eds), Clinical insights: chronic pain. Future Medicine Ltd., London, pp 5–17
4. Borrell-Carrio F, Suchman AL, Epstein RM (2004) The Biopsychosocial model 25 years later: principles, practice, and scientific inquiry. Ann Fam Med 2:576–582. https://doi.org/10.1370/afm.245
5. Engel GL (1979) The Biopsychosocial model and the education of health professionals. Gen Hosp Psychiatry 1:156–165. https://doi.org/10.1016/0163-8343(79)90062-8
6. Margalit APA, Glick SM, Benbassat J et al (2004) Effect of a biopsychosocial approach on patient satisfaction and patterns of care. J Gen Intern Med 19(5Pt2):485–491. https://doi.org/10.1111/j.1525-1497.2004.30059.x
7. Matri S (2008) Holistic medicine and the western medical tradition. Virtual Mentor 10 (3):177–180
8. Center for Disease Control (2015), Fibromyalgia. Atlanta: Centers for Disease Control
9. Katz JD, Mamyrova G, Guzhva O et al (2010) Gender bias in diagnosing fibromyalgia. Gend Med 7(1):19–27. https://doi.org/10.1016/j.genm.2010.01.003
10. González E, Elorza J, Failde I (2010) Fibromyalgia and psychiatric comorbidity: Their effect on the quality of life patients. Actas Esp Psiquiatr 38(5):295–300
11. Arnold LM, Crofford LJ, Mease PJ et al (2008) Patient perspectives on the impact of fibromyalgia. Patient Educ Couns 73(1):114–120. https://doi.org/10.1016/j.pec.2008.06.005

12. Wolfe F, Clauw DJ, Fitzcharles MA et al (2011) Fibromyalgia criteria and severity scales for clinical and epidemiological studies: a modification of the ACR preliminary diagnostic criteria for fibromyalgia. J Rheumatol 38(6):1113–1122. https://doi.org/10.3899/jrheum.100594
13. Macfarlane GJ, Kronisch C, Dean LE et al (2016) EULAR revised recommendations for the management of fibromyalgia. Ann Rheum Dis. https://doi.org/10.1136/annrheumdis-2016-209724
14. Heron KE, Smyth JM (2010) Ecological momentary interventions: Incorporating mobile technology into psychosocial and health behaviour treatments. Br J Health Psychol 15(Pt 1):1–39. https://doi.org/10.1348/135910709X466063
15. Kool MB, Greenen R (2012) Loneliness in patients with rheumatic diseases: The significance of invalidation and lack of social support. J Psychol 146(1–2):229–241
16. Daraz L et al (2011) The quality of websites addressing fibromyalgia: an assessment of quality and readability using standardised tools. BMJ Open 1(1):e000152. https://doi.org/10.1136/bmjopen-2011-000152
17. Friedberg F, Williams DA, Collinge W (2012) Lifestyle-oriented non-pharmacological treatments for fibromyalgia: A clinical overview and applications with home-based technologies. J Pain Res. 5:425–435. https://doi.org/10.2147/JPR.S35199
18. Williams DA, Kuper D, Segar M et al (2010) Internet-enhanced management of fibromyalgia: a randomized controlled trial. Pain 151(3):694–702. https://doi.org/10.1016/j.pain.2010.08.034
19. Menga G, Ing S, Khan O et al (2014) Fibromyalgia: can online cognitive behavioral therapy help? Ochsner J 14(3):343–349
20. Collinge W, Soltysik RC (2010) Yarnold PR (2010) An internet-based intervention for fibromyalgia self-management: initial design and alpha test. Optim Data Anal 1(1):163–175
21. Collinge W, Yarnold P, Soltysik R (2013) Fibromyalgia symptom reduction by online behavioral self-monitoring, longitudinal single subject analysis and automated delivery of individualized guidance. N Am J Med Sci 5(9):546–553. https://doi.org/10.4103/1947-2714.118920
22. Mortensen J, Kristensen LQ, Brooks EP et al (2015) Women with fibromyalgia's experience with three motion-controlled video game consoles and indicators of symptom severity and performance of activities of daily living. Disabil Rehabil Assist Technol 10(1):61–66. https://doi.org/10.3109/17483107.2013.836687
23. Garcia-Palacios A, Herrero R, Belmonte MA et al (2014) Ecological momentary assessment for chronic pain in fibromyalgia using a smartphone: a randomized crossover study. Eur J Pain 18(6):862–872
24. Vanderboom CE, Vincent A, Luedtke CA et al (2014) Feasibility of interactive technology for symptom monitoring in patients with fibromyalgia. Pain Manag Nurs 15(3):557–564. https://doi.org/10.1016/j.pmn.2012.12.001
25. Wells AF, Arnold LM, Curtis CE et al (2013) Integrating health information technology and electronic health records into the management of fibromyalgia. Postgrad Med 125(4):70–77. https://doi.org/10.3810/pgm.2013.07.2678
26. Kayyali HA, Weimer S, Frederick C et al (2008) Remotely attended home monitoring of sleep disorders. Telemed J E Health 14(4):371–374. https://doi.org/10.1089/tmj.2007.0058
27. De Jonckheere J, Ibarissene I, Flocteil M et al (2014) A smartphone based cardiac coherence biofeedback system. Conf Proc IEEE Eng Med Biol Soc 2014:4791–4794. https://doi.org/10.1109/EMBC.2014.6944695

Physical Activity and Women's Mental Health

Eluana Gomes, Raquel Costa, Tânia Bastos, Michel Probst and Rui Corredeira

Abstract Women are twice as likely as men to develop certain mental health conditions such as depression, eating disorders, and anxiety disorders. A multimodal care, including psychotherapies as adjuncts to antipsychotic medications, is acknowledged to be crucial in teaching individual strategies and providing patients with tools to deal with these illnesses. In this scenario, physical activity has become increasingly relevant to promote physical and mental health in women with mental illness. However, it appears that women have unique experiences, risks, and needs that must be taken into account for the treatment strategies. For this reason, the provision of rehabilitation for women with mental illness has been challenging for the mental health systems reform. This underlines the importance of understanding the modern scientific and nonscientific literature about the link between physical activity and mental health in women. This chapter will present research on the relationship between physical activity and mental illness, trying to list female particularities and the advantages of exercise for their health.

Keywords Mental illness · Physical activity programs
Health · Women

E. Gomes (✉) · R. Costa · R. Corredeira
Research Center in Physical Activity, Health and Leisure (CIAFEL),
Faculty of Sport, University of Porto, Porto, Portugal
e-mail: aeluana@yahoo.com.br

T. Bastos
Center of Research, Education, Innovation and Intervention in Sport (CIFI2D),
Faculty of Sport, University of Porto, Porto, Portugal

T. Bastos
Research Center in Sports Sciences, Health Sciences and Human Development (CIDESD),
ISMAI - Instituto Superior Da Maia, Porto, Portugal

M. Probst
Research Group for Adapted Physical Activity and Psychomotor Rehabilitation,
Faculty of Kinesiology and Rehabilitation Sciences, KU Leuven, Leuven, Belgium

S. Brandão et al. (eds.), *Women's Health and Biomechanics*,
Lecture Notes in Computational Vision and Biomechanics 29,
https://doi.org/10.1007/978-3-319-71574-2_2

Highlights

- The rehabilitation for women with mental illness has been challenging for the mental health systems.
- Physical activity can act as a complementary therapy and become relevant to promote physical and mental health in women with mental illness.
- Women with mental illness presented low physical activity levels that can be explained due to the lack of sufficient motivation.
- Little is still known about the women's preferences towards physical activity and also their expectation and goals when practicing.
- The chapter presents the current state-of-the-art about the relationship between physical activity and mental illness focusing on the women's characteristics.

1 Introduction

Health in its broader sense from 1948 can be seen as "a state of complete physical, mental, and social well-being and not merely the absence of disease or infirmity" [1]. Regular physical activity has significant relevance when promoting the physical and mental health of the population [2]. However, at least 60% of the global population fails to achieve the minimum recommendation for health-related physical fitness (i.e., 30 min of daily moderate-intensity physical activity) [3]. Physical inactivity is generally more prevalent among girls and women than their male counterparts [4].

Some diseases influence the death and disability rates for women around the world. For example, cardiovascular diseases account for one-third of deaths among women around the world and half of all deaths are of women over 50 years old in developing countries. *Diabetes mellitus* affects more than 70 million women in the world and its prevalence is expected to double by 2025. Osteoporosis is the most prevalent disease in postmenopausal women and breast cancer is the most commonly diagnosed cancer in women [4]. In this context, regular physical activity is an important tool to promote health in women. Some benefits related to the practice of physical activity in women include: reducing the risk of death from coronary heart disease and of developing high blood pressure; reducing the risk of developing colon cancer; reducing the risk of developing diabetes; helping to maintain healthy bones, muscles, and joints; helping control weight, building lean muscle; reducing body fat; helping control joint swelling and the pain associated with arthritis; and enhancing the effect of estrogen replacement therapy in decreasing bone loss after menopause [5].

Mental illness is one of the health problems in the twenty-first century. It may not, in itself, be fatal but causes extensive disability in rich and poor countries alike, and has been increasing. It comprises a broad range of problems, with different symptoms [6] that are generally characterized by some combination of abnormal thoughts, emotions, behavior, and relationships [7]. The burden of mental illness continues to grow, with significant impact on health and economic consequences in all countries of the world [8]. According to the World Health Organization [9], one in four people in the world will be affected by mental illness at some point in their lives. Mental illnesses are among the leading causes of ill-health and disability worldwide. Individuals with mental illness are at high risk of chronic disease, including diabetes, hyperlipidemia, and cardiovascular disease [10]. Individuals with mental illness experience loss of energy, interest, and motivation towards daily life activities. It is known that general fatigue is directly related to a more sedentary lifestyle and poor physical health [11]. In addition, these individuals have unhealthy lifestyle patterns, which include poor diet and high rates of cigarette smoking [12].

Regarding women with mental illness, the Substance Abuse and Mental Health Services Administration [13] reports that approximately 23.8% of American women have experienced a diagnosable mental health disorder, compared to the estimated 15.6% of men who have mental illness. In this context, the biological factors seem to play an important role in mental illness. For example, women have lower serotonin levels than men and also process the chemical at slower rates, which can contribute to fluctuations in mood. This is why females are generally more predisposed to hormonal fluctuations [14]. Furthermore, women have historically been regarded as the subordinate gender, caregivers for their children and the elderly [15]. However, in the last decades, women have had a chance to secure managerial posts or reach high-ranking positions in the careers [16]. This stress scenario may lead to depression and panic attacks in women [17, 18]. Anxiety disorders, bipolar disorder, depression, postnatal depression, eating disorders, and schizophrenia are some of the common mental health illnesses in women [18]. Depression, organic brain syndromes, and dementias are most common in elderly women [19].

It appears that women have unique experiences, risks, and needs that must be considered for treatment strategies [20]. For this reason, the provision of rehabilitation for women with mental illness has been challenging for the mental health systems reform. Women are particularly exposed to some of the factors that increase the risk of poor mental health because of the role and status they typically have in society (e.g., physical and sexual abuse) [21].

Physical activity has been shown to enhance the effectiveness of psychological therapies and to have a role in improving the quality of life and symptom management for individuals with a wide range of mental health problems [22]. Regular physical activity is associated with several health benefits for individuals with mental illness, namely: (i) to improve quality of sleep; (ii) to increase interest in sex; (iii) to increase endurance; (iv) to promote stress relief; (v) to improve mood; (vi) to increase energy and stamina; (vii) to reduce tiredness which can increase mental alertness; (viii) to control weight; (ix) to reduce cholesterol and; (x) to improve cardiovascular fitness [10]. It is important to mention that physical activity can also have a positive impact on this population's self-esteem and social withdrawal [23].

Physical activity for women with mental illness reduces stress, anxiety, and mood state [24]. Specifically, physical activity helps to decrease depressive symptoms among women with depression [25] and postnatal depression [26], and also to improve body composition of women with eating disorders [27]. These health benefits should be emphasized and reinforced by every mental health professional to their patients.

This unfavorable scenario for women with mental health underlines the importance of understanding the modern scientific and nonscientific literature about the link between the physical activity and mental health in women. This chapter will present the current state-of-the-art research on the relation between physical activity and mental illness, trying to list female gender particularities and the advantages of being physically active for the global definition of health.

2 Gender Differences in Mental Illness

It is known that both men and women with mental illness face greater medical challenges when compared to individuals without mental illness [28]. Specifically, experience high rates of multiple or comorbid medical problems such as cardiovascular, endocrine, and respiratory illnesses [20, 28]. This is due to a history of trauma (e.g., sexual abuse, domestic violence), barriers to treatment of physical illness (e.g., poverty, lack of insurance; misdiagnosis, or underdiagnosis), lifestyle choices (e.g., high smoking prevalence and substance misuse), the effects of medications (e.g., obesity and *Diabetes Mellitus* related to psychiatric medications), and consequences of the illness itself (e.g., neglect of personal care) [20]. In addition, gender differences are observed in the prevalence and symptoms. According to the National Institute of Mental Health [29], both women and men are influenced by biological and psychosocial factors that may lead to differences in mental illness. Women show higher rates of mood and anxiety disorders while men exhibit higher rates of antisocial personality and substance use disorders [30].

It has also been reported that women are 48% more likely than men to use any psychotropic medication [31]. Therefore, they also experience psychotropic medication-related problems (e.g., weight gain, amenorrhea, sexual dysfunction) and long-term medication effects (e.g., increased rates of breast cancer, osteoporosis, immune disorders, and cardiac complications) [28]. The Mental Health Foundation [21] reports that women are more likely to be treated for a mental health problem when compared to men (29% vs. 17%, respectively). This reflects women's greater willingness to acknowledge that they are troubled and get support. It is also known that women are more likely to seek help from and disclose mental health problems to their primary health care physician [32]. On the other hand, men are more likely to seek specialist mental health care and, consequently, they are the principal users of inpatient care [33].

3 The Rehabilitation Process in Mental Illness

In the past, severe mental illness was generally considered chronic and incapacitating diseases that worsened over the life course [34]. This belief has discouraged patients from engaging in normative activities, such as education, employment, childrearing, and independent living [35]. Patients frequently developed service-dependent lifestyles, involving lengthy institutionalization, heavy medication, sheltered activities, and supervised housing [36].

In the mid-1950s, and beginning in England, United States of America, and in Continental Europe, the World Health Organization promoted an international *consensus* about the need of a thorough change in psychiatric care and new policy strategies for mental health. This *consensus* aimed at overcoming the old asylum-based system of care and establishing new community-oriented therapeutic approaches [37]. The rehabilitation process appears as a multidimensional construct that includes five dimensions: (i) clinical (e.g., medical care and psychotropic medication); (ii) existential (e.g., religion and spirituality); (iii) functional (e.g., employment and education); (iv) physical (e.g., diet, exercise and smoking) and; v) social (e.g., family and social activity) [38]. Therefore, a multimodal care, including psychotherapies as adjuncts to medications, is acknowledged to be crucial in teaching individual strategies and to give patients tools to deal with stress and unhealthy thoughts and behaviors [18]. Nowadays, there are multiple approaches to treatment such as counseling, cognitive behavioral therapy, social skills training, family psycho-education, assertive community treatment, and supported employment, which may include multidisciplinary mental health professionals. The effectiveness of physical therapy interventions within the multidisciplinary treatment has become increasingly relevant. This is why it is crucial to promote healthy lifestyles among individuals living with serious mental illness as an integral part of the rehabilitation process.

However, the provision of rehabilitation for women with mental illness has been, and still is, one of the major challenges for mental health systems reform [39]. Psychosocial rehabilitation must be designed to answer the specific needs of women. Therefore, treatments need to be sensitive to and reflect gender differences [21].

4 The Relevance of Physical Activity in Women with Mental Illness

Nowadays, physical activity can act as a complementary therapy or as a supplement to bio-psychotherapeutic approaches (behavior, cognitive, or psychodynamic therapy) for different diagnoses. It incorporates medical, psychological, pedagogic, kinesiological, and rehabilitative components [22]. For these reasons, researchers recommend the inclusion of physical activity as an adjunct treatment in psychiatric rehabilitation [11, 23, 25, 40].

It should be noted that most physical activity programs are developed for men and women with different types of mental illness. The literature reported positive effects of physical activity programs in both men and women with schizophrenia [41, 42], bipolar disorder [43], post-traumatic stress disorder [40], anxiety disorders [44], obsessive-compulsive disorder [45], panic disorder [46], substance use disorders and dementia/mild cognitive [47]. A few researchers have developed physical activity programs specifically for women with mental illness and positive effects have also been reported in depressive disorders [25, 48, 49], postnatal depression [26, 50], and eating disorders [27, 51].

The physical activity guidelines for individuals with mental illness are not clear in the literature. Comparative analyses suggest physical activity programs for individuals with mental illness are often difficult due to variations in the level of severity of the disease, methodological weaknesses (e.g., small sample size) and the characteristics of the intervention delivered (i.e., frequency, type, and duration).

Therefore, it is difficult to reach a *consensus* regarding the best physical activity program in the context of mental illness. Nevertheless, there are some guidelines for physical activity for some specific mental illnesses. For example, the National Institute for Health and Clinical Excellence [52] has recommended structured, supervised exercise programs for individuals with depression (i.e., mild to moderate) three times a week (45 min to 1 h) over a period of 10–14 weeks, and based on low-intensity. The American Psychiatric Association [53] has recommended physical activity programs for individuals with panic disorders with at least 4 days a week of aerobic exercise (walking for 60 min or running for 20–30 min). Vancampfort et al. [54] suggested that individuals with schizophrenia should do at least 150 min a week of moderate intensity, or 75 min of moderate to vigorous aerobic exercises. The activities should be carried out in instances of at least 10 min, and at least 3 days a week.

The guidelines for physical activity programs, specifically designed for women with mental illness are even scarcer but we could identify some studies in this context, as shown next. Lin et al. [55] applied yoga and aerobic exercise (walking and cycling) in female patients with early psychosis, three times a week with of 60 min per session.

Usually, the physical activity programs for women with depression have three [25, 49] to four sessions a week [48]. These sessions range from 30 [48] to 50 min each [25]. The types of exercise offered were aerobic [25], treadmill aerobic [48, 49] and home-based exercise [56].

Armstrong and Edwards developed different physical activity programs for women with postnatal depression ranging in frequency between twice-a-week [26] to three times a week [50]. The duration of the sessions would range from 30 min [50] to 40 min [26]. The type of exercise involved was walking and pram walking group [26, 50]. In previously mentioned studies, the researchers highlighted the importance of social support throughout the intervention.

In general, the studies that developed programs for women with Alzheimer's disease reported a frequency average of three times a week [57, 58], the sessions would last between 40 [59] and 60 min per session [57, 58]. Stretching, muscular resistance, rhythmic activities, and balance [57, 58] were the most common activities implemented.

Women with eating disorders have engaged in physical activity programs with a frequency of three times per week [27, 51] and with duration of 60 min per session [27, 51]. The types of exercise implemented were resistance training [51], volleyball, soccer, and walking [27]. In addition, Bentley et al. [60] recommended a weight restoration program for individuals with eating disorders with a body mass index of 15/17 and above. Exercising in groups is strongly advised for this population, as it tends to limit uncontrolled activity, is time-limited and has the added benefit of social interaction.

In general, the studies for population with mental illness perform exercises with moderate to vigorous intensities. Different instruments such as Borg Scale, Visual Analogue Scale, pedometers, heart rate monitors, accelerometers can be applied to monitor exercise intensity [25, 27, 42, 43]. Physical activity programs for a population with mental illness can be delivered by different professionals, such as nurses [61], exercise physiologists [27, 62], exercise therapists [49] and physical education teachers [25, 42]. Some programs are developed by combining counseling on nutrition and healthy lifestyles [20, 40, 44].

Long-term adherence to physical activity is essential for maintaining health benefits. In the general population, adherence to physical activity programs decreases after 6 months to half of the original number of participants [63]. Therefore, it would be unreasonable to expect better from programs for individuals with mental illnesses, who have additional barriers to regular attendance (e.g., illness exacerbation, issues surrounding autonomy/independence, and increased motivational problems) [23]. In this scenario, using motivational strategies to increase their adherence is important to enhance their success. The adherence strategies more frequently reported in the literature are: (i) to organize the exercises with small groups of participants; (ii) to provide adequate trained support; (iii) to continuously encourage and give positive feedback throughout the sessions; (iv) to make telephone calls to participants who missed physical activity sessions; and (v) to establish a trusting relationship between participants and the research team through discussions focused on physical activity and healthy lifestyles, or by sharing personal experiences [23, 42, 64–66]. Lastly, researchers pointed that the individuals that are living near and have easy access (i.e., central location, near public transport) to a mental health service with physical activity facilities would have better adherence [42, 54].

Finally, getting to know about the type of exercise that women with mental illness prefer can be another tool to improve the adherence to physical activity programs. Evidence suggests that social support is consistently related to physical activity adherence for healthy women. For example, younger university students and adult women prefer structured exercise classes, admitting a preference for group-based activities where social interaction is foundational [67, 68].

However, women with social physique anxiety (i.e., anxiety that people experience in response to others' evaluations of their physique) tend to exercise alone rather than in a group context, to have more favorable attitudes toward group exercise settings that do not emphasize the physique (i.e., where less revealing clothing is worn) and to experience increased state anxiety during exercise in public settings [67].

Overall the importance of regular physical activity for health and for the improvement of the quality of life of individuals with mental illness has been widely accepted in the literature. Physical activity and lifestyle counseling are important tools for the rehabilitation process aiming to improve the functionality and the independence of this group.

5 Conclusion

Most physical activity programs available for individuals with mental illness cover both men and women. However, the participation of women in physical activity programs is low. It seems that less attention is being given to women's preferences towards physical activity to physical, psychological, and social characteristics. Therefore, it is necessary to promote additional research to assess the effects of physical activity in women with mental illness. Individuals with mental illness have low motivation levels for the practice of physical activity and this situation is even worse in women. Little is still known about the process of initiating women in physical activity, namely what are the reasons they engage in physical activity that make them be physically active. Future studies should try to understand their previous experiences of practice and also their expectation and goals when practicing physical activity.

The literature recognizes that it is necessary to implement motivational strategies to increase attendance rates on physical activities but we could not identify specific strategies for women with mental illness. Using the general population as a reference, it is possible to suggest that physical activity should be attractive, exciting, and framed in the preferences of women towards physical activity. The use of a group-based intervention, where social interaction prevails also seems to be an appealing context for women. However, it is necessary to confirm whether these preferences are also the ones of women with mental illness or whether they have different preferences.

In this context, the role of physical activity professionals is of high importance. These technicians have to know the specific characteristics and symptoms of each disease and consequently the specific needs of both genders to design proper interventions. All these parameters are essential to reverse the weak scenario with regard to opportunities and rates of practice of physical activity among women with mental illness.

References

1. World Health Organization (1946) Constitution of the World Health Organization as adopted by the International Health Conference, New York, (Official Records of the World Health Organization, no 2, p 100
2. World Health Organization (2006) Constitution of the World Health Organization—Basic Documents, Forty-fifth edn, Supplement
3. World Health Organization (2002) World Health Report—Reducing risks, promoting healthy life. Geneva World Health Organization
4. World Health Organization (2016) Physical Activity and Women (s.d.). http://www.who.int/dietphysicalactivity/factsheet_women/en/. Accessed 2 Apr 2016
5. Centers for Disease Control and Prevention (1999) Physical Activity and Health: A Report of the Surgeon General—Women. http://www.cdc.gov/nccdphp/sgr/women.htm. Accessed 4 Apr 2016
6. Brundtland GH, Mental health in the 21st century. Bull World Health Organ 78(4) (2000)
7. World Health Organization (2016) Mental disorders. http://www.who.int/topics/mental_disorders/en/. Accessed 6 Apr 2016
8. World Health Organization (2015). Mental disorders. http://www.who.int/mediacentre/factsheets/fs396/en/. Accessed 4 Apr 2016
9. World Health Organization (2001) Mental disorders affect one in four people. http://www.who.int/whr/2001/media_centre/press_release/en/. Accessed 5 Dec 2015
10. Sharma A, Madaan V, Petty FD (2006) Exercise for mental health. Prim Care Companion J Clin Psychiatry 8(2):106
11. Vancampfort D, Sweers K, Probst M et al (2011) Quality assessment of physical activity within clinical guidelines for the prevention and treatment of cardiometabolic risk in people with schizophrenia. Commun Mental Health J 47(6):703–710. https://doi.org/10.1007/s10597-011-9431-8
12. Compton MT, Daumit GL, Druss BG (2006) Cigarette smoking and overweight/obesity among individuals with serious mental illnesses: a preventive perspective. Harv Rev Psychiatry 14:212–222
13. Substance Abuse and Mental Health Services Administration (SAMHSA, 2010). Circles of care. http://store.samhsa.gov/shin/content//SAM10-186/SAM10-186.pdf. Accessed 2 Apr 2016
14. Nishizawa S, Benkelfat C, Young SN et al (1997) Differences between males and females in rates of serotonin synthesis in human brain. Proc Natl Acad Sci 94(10):5308–5313
15. Brody EM, Johnsen PT, Fulcomer MC (1984) What should adult children do for elderly parents? Opinions and preferences of three generations of women. J Gerontol 39(6):736–746
16. Mishler, S (1975) Barriers to the career development of women. In: Osipow, Samuel H. (Ed), Emerging woman: career analysis and outlooks. Oxford, England: Charles E. Merrill, p 161
17. Wood C, Cano-Vindel A, Martín Salguero J (2015) A multi-factor model of panic disorder: results of a preliminary study integrating the role of perfectionism, stress, physiological anxiety and anxiety sensitivity. Anales Psicologia 31(2):481–487
18. National Institute of Mental Health (s.d.). http://www.nimh.nih.gov/health/topics/women-and-mental-health/index.shtml. Accessed 6 Apr 2016
19. World Health Organization (2016) Gender disparities and mental health: the facts (s.d.). http://www.who.int/mental_health/prevention/genderwomen/en/. Accessed 6 Apr 2016
20. Vandiver V (2007) Health promotion as brief treatment: strategies for women with co-morbid health and mental health conditions. Brief Treat Crisis Intervention 7:161–175
21. Mental Health Foundation (2015) Women and Mental Health (s.d.). http://www.mentalhealth.org.uk/help-information/mental-health-a-z/W/women/. Accessed 15 Dec 2015
22. Probst M, Carraro A (eds) (2013) Physical activity and mental health, a practice oriented approach. Milano, New York: EdiErmes

23. Richardson C, Faulkner G, McDevitt J et al (2005) Integrating physical activity into mental health services for persons with serious mental illness. Psychiatr Serv 56(3):324–331
24. Rejeski WJ, Thompson A, Brubaker PH et al (1992) Acute exercise: buffering psychosocial stress responses in women. Health Psychol 11(6):355–362
25. Carneiro L, Fonseca AM, Vieira-Coelho MA et al (2015) Effects of structured exercise and pharmacotherapy vs. pharmacotherapy for adults with depressive symptoms: a randomized clinical trial. J Psychiat Res 71:48–55. https://doi.org/10.1016/j.jpsychires.2015.09.007
26. Armstrong K, Edwards H (2004) The effectiveness of a pram-walking exercise programme in reducing depressive symptomatology for postnatal women. Int J Nurs Pract 10:177–194
27. Bratland-Sanda S, Sundgot-Borgen J, Rø Ø et al (2010) Physical activity and exercise dependence during inpatient treatment of longstanding eating disorders: an exploratory study of excessive and non-excessive exercisers. Int J Eat Disord 43(3):266–273. https://doi.org/10.1002/eat.20769
28. Perese E, Perese K (2003) Health problems of women with severe mental illness. J Am Acad Nurse Pract 15(5):212–219. https://doi.org/10.1111/j.1745-7599.2003.tb00361.x
29. National Institute of Mental Health. Psychotherapies (s.d.). http://www.nimh.nih.gov/health/topics/psychotherapies/index.shtml. Accessed 4 Dec 2015
30. Eaton N et al (2912) An invariant dimensional liability model of gender differences in mental disorder prevalence: evidence from a national sample. J Abnorm Psychol 121(1):282–288
31. Simoni-Wastila L (2000) The use of abusable prescription drugs: the role of gender. J Womens Health Gend Based Med 9(3):289–297. https://doi.org/10.1089/152460900318470
32. National Institute of Health. Women and Mental Health (s.d.). http://www.nimh.nih.gov/health/statistics/prevalence/use-of-mental-health-services-and-treatment-among-adults.shtml. Accessed 16 Dec 2015
33. World Health Organization. The global burden of disease: 2004 update. Geneva: World Health Organization
34. Bellack AS (2006) Scientific and consumer models of recovery in schizophrenia: concordance, contrasts, and implications. Schizophr Bull 32(3):432–442. https://doi.org/10.1093/schbul/sbj044
35. Shorter E (1997) A history of psychiatry: from the era of the asylum to the age of Prozac. Wiley, New York (NY)
36. Davidson L (2003) Living outside mental illness: qualitative studies of recovery in schizophrenia. New York University Press, New York
37. Freeman HL, Fryers T, Henderson JH (1985). Mental health services in Europe: 10 years on. Copenhagen: WHO Regional Office for Europe
38. Whitley R, Drake RE (2010) Recovery: a dimensional approach. Psychiatr Serv 61(12):1248–1250. https://doi.org/10.1176/ps.2010.61.12.1248
39. Chatterjee R, Hashim U (2015) Rehabilitation of mentally ill women. Indian J Psychiatry 57(6):345–353. https://doi.org/10.4103/0019-5545.161503
40. Rosenbaum S, Vancampfort D, Steel Z et al (2015) Physical activity in the treatment of post-traumatic stress disorder: a systematic review and meta-analysis. Psychiatry Res 230:130–136. https://doi.org/10.1016/j.psychres.2015.10.017
41. Vancampfort D, Knapen J, Probst M et al (2012) A systematic review of correlates of physical activity in patients with schizophrenia. Acta Psychiatr Scand 125(5):352–362. https://doi.org/10.1111/j.1600-0447.2011.01814
42. Gomes E, Bastos T, Probst M et al (2014) Effects of a group physical activity program on physical fitness and quality of life in individuals with schizophrenia. Mental Health Phys Activity 7(3):155–162. https://doi.org/10.1016/j.mhpa.2014.07.002
43. Sylvia LG, Salcedo S, Bernstein EE et al (2013) Nutrition, exercise, and wellness treatment in bipolar disorder: proof of concept for a consolidated intervention. Int J Bipolar Disord 1(1):24. https://doi.org/10.1186/2194-7511-1-2
44. Jayakody K, Gunadasa S, Hosker C (2014) Exercise for anxiety disorders: systematic review. Br J Sports Med 48(3):187–196. https://doi.org/10.1136/bjsports-2012-091287

45. Readdy T, Ebbeck V (2013) Obsessive-compulsive disorder symptoms and correlates in community exercisers. Psychol Sport Exerc 14:316–322
46. Esquivel G, Díaz-Galvis J, Schruers K et al (2008) Acute exercise reduces the effects of a 35% CO2 challenge in patients with panic disorder. J Affect Disord 107(1–3):217–220. https://doi.org/10.1016/j.jad.2007.07.022
47. Zschucke E, Gaudlitz K, Ströhle A (2013) Exercise and physical activity in mental disorders: clinical and experimental evidence. J Prev Med Public Health 46:2–21. https://doi.org/10.3961/jpmph.2013.46.S.S12
48. Chu I-H, Kirbyc TE, Emeryd CF et al (2009) Effect of exercise intensity on depressive symptoms in women. Ment Health Phys Act 2:37–43. https://doi.org/10.1016/j.mhpa.2009.01.001
49. Callaghan P, Khalil E, Morres I et al (2011) Pragmatic randomised controlled trial of preferred intensity exercise in women living with depression. BMC Public Health 11:465. https://doi.org/10.1186/1471-2458-11-465
50. Armstrong K, Edwards H (2003) The effects of exercise and social support on mothers reporting depressive symptoms: A pilot randomized controlled trial. Int J Ment Health Nurs 23:130–138
51. Chantler I, Szabo CP, Green K (2006) Muscular strength changes in hospitalized anorexic patients after an eight week resistance training program. Int J Sports Med 27(8):660–665
52. National Institute for Health and Clinical Excellence (2009) Depression: the Treatment and Management of Depression in Adults (Update). http://www.nice.org.uk/guidance/CG90. Accessed 8 Apr 2016
53. American Psychiatric Association (2009) Treatment of Patients with Panic Disorder (Guideline). American Psychiatric Association, 2nd edn
54. Vancampfort D, De Hert M, Skjerven LH et al (2012) International organization of physical therapy in mental health consensus on physical activity within multidisciplinary rehabilitation programmes for minimising cardiometabolic risk in patients with schizophrenia. Disabil Rehabilit 34(1):1–12. https://doi.org/10.3109/09638288.2011.587090
55. Lin J, Chan SK, Lee EH et al (2015) Aerobic exercise and yoga improve neurocognitive function in women with early psychosis. NPJ Schizophr, p 15047. https://doi.org/10.1038/npjschz.2015.47
56. Da Costa D, Lowensteyn I, Abrahamowicz M et al (2009) A randomized clinical trial of exercise to alleviate postpartum depressed mood. J Psychosom Obst Gyn 30(3):191–200. https://doi.org/10.1080/01674820903212136
57. Nascimento CM, Teixeira CV, Gobbi LT et al (2012) A controlled clinical trial on the effects of exercise on neuropsychiatric disorders and instrumental activities in women with Alzheimer's disease. Revis Brasile de Fisioter 16(3):197–204
58. Groppo HS, Nascimento CMC, Stelaa F et al (2010) Efeitos de um programa de atividade física sobre os sintomas depressivos e a qualidade de vida de idosos com demência de Alzheimer. Rev Bras Educ Fís Esporte 26(4):543–551
59. Namazi KH, Zadorozny CA, Gwinnup PB et al (1995) The influences of physical activity on patterns of sleep behavior of patients with Alzheimer's disease. Int J Aging Hum Dev 40 (2):145–153
60. Bentley J, Caddy P, Hammond L et al (2011) Physiotherapy guidance notes for exercise and physical activity in adult patients with anorexia and bulimia nervosa.http://www.rcpsych.ac.uk/pdf/eating%20disorders%20physiotherapy%20june%202011.pdf
61. Happell B, Plantania-Phung C, Scott D (2011) Placing physical activity in mental health care: a leadership role for mental health nurses. Int J Ment Heal Nurs 20(5):310–318
62. Heggelund J, Nilsberg GE, Hoff J et al (2011) Effects of high aerobic intensity training in patients with schizophrenia: a controlled trial. Nordic J Psychia 65(4):269–275. https://doi.org/10.3109/08039488.2011.560278
63. Bingham B (2009) Physical Activity and Mental Health Literature Review. Canadian Mental Health Association, Ontario, YMCA Ontario, York University, Ontario Ministry of Health Promotion, pp 1–16

64. Beebe LH, Harris RF (2012) Using pedometers to document physical activity in patients with schizophrenia spectrum disorders a feasibility study. J Psychosoc Nurs Ment Health Serv 50 (2):44–49. https://doi.org/10.3928/02793695-20120112-02
65. Bernard P, Ninot G (2012) Bénéfices des activités physiques adaptéesdans la prise en charge de la schizophrénie: revue systématique de la littérature. L'Encéphale 38(4):280–287. https://doi.org/10.1016/j.encep.2011.07.001
66. Warren KR, Ball MP, Feldman S et al (2011) Exercise program adherence using a 5-KiPalntanialometer (5 K) event as an achievable goal in people with schizophrenia. Biol Res Nurs 13(4):383–390. https://doi.org/10.1177/1099800410393272
67. Burke SM, Carrona AV, Eys MA (2006) Physical activity context: Preferences of university students. Psychol Sport Exerc 7(1):1–13. https://doi.org/10.1016/j.psychsport.2005.03.002
68. Tuakli-Wosornu YA, Rowan M, Gittelsohn J (2014) Perceptions of physical activity, activity preferences and health among a group of adult women in Urban Ghana: a pilot study. Ghana Med J 48(1):3–13

The Role of Ultrasound Imaging of Musculotendinous Structures in the Elderly Population

Luis Carrão, Rute Santos, Margarida Espanha, Paulo Armada-da-Silva, Ana Amaral, Sandra Amado, Henrique A. Almeida, Paula Faria and António Veloso

Abstract Ultrasound (US) is a noninvasive and real-time method that allows the evaluating muscles and tendons. The enhanced echo-intensity (EI) on ultrasonography images of skeletal muscle is believed to reflect changes in muscle quality (MQ), and these changes accompany aging. Also related to aging, and that may more severely affect women than men, is the well-known loss of skeletal muscle mass. Often associated with the accumulation of connective tissues (e.g., adipose), it affects muscle strength and MQ and causes functional impairment. This chapter demonstrates the potential use of US imaging for assessing muscle changes associated with aging and functional decline.

Keywords Ultrasound · Echo-intensity · Musculoskeletal system · Aging

---

L. Carrão (✉) · R. Santos · M. Espanha · P. Armada-da-Silva · S. Amado · A. Veloso
CIPER – LBMF, Faculty of Human Kinetics, University of Lisbon, Lisbon, Portugal
e-mail: lmcctfisio@gmail.com

L. Carrão
ESSLEI, Polytechnic Institute of Leiria, Leiria, Portugal

R. Santos · A. Amaral
ESTeS Coimbra, Polytechnic Institute of Coimbra, Coimbra, Portugal

S. Amado · P. Faria
CDRSP, Polytechnic Institute of Leiria, Leiria, Portugal

H. A. Almeida
School of Technology and Management Polytechnic Institute of Leiria, Leiria, Portugal

P. Faria
ESTG, Polytechnic Institute of Leiria, Leiria, Portugal

S. Brandão et al. (eds.), *Women's Health and Biomechanics*,
Lecture Notes in Computational Vision and Biomechanics 29,
https://doi.org/10.1007/978-3-319-71574-2_3

Highlights

- Ultrasound Imaging may be used to assess the aging process related to the musculoskeletal system.
- Muscles and tendons can be evaluated regarding their echo-intensity.
- Ultrasound Imaging features are related to functional decline.

1 Introduction

Aging is a continuous process associated with significant structural and functional changes in the body, and it is one of the most significant trends of the century [1]. Several changes occur during aging, especially in the musculoskeletal system [2, 3]. Loss of muscle mass accompanied by accumulation of connective tissues (e.g., fat) affects the strength and quality of muscles (describe by the physiological functional capacity of muscle tissue) [4], impairing gait and balance, increasing the risk of falling and leading to loss of physical independence [1, 3, 5]. Changes in muscle architecture and loss of stiffness of tendons and ligaments are also associated with aging [6].

The impact of aging in quality-of-life has been studied to evaluate the elderly health and the implementation of strategies and intervention programs, in order to delay or decrease such effects [5]. According to the World Health Organization, regular physical activity programs must be implemented in order to reduce the aging effects, providing a healthier lifestyle. According to the literature, physical activity in the elderly population has physical, functional, and physiological benefits and helps preventing diseases resulting from the aging process [6].

In aging as well in osteoarthritis (OA) patients, the decrement in strength is greater than would be expected due to disuse atrophy, consequent to a reduction in loading of the extremity and reduction in muscle size [7]. This suggests that modifications in the intrinsic characteristics of muscles, reflecting a diminished muscle quality (MQ), may also contribute to impaired muscle function in the elderly and of those with knee OA [8], which is a major clinical problem affecting more women than men [9].

In the elderly population, the most prevalent joint disease is OA and knee OA is a major cause of disability [10, 11]. Symptomatic knee OA affects roughly 12% of the people beyond 60 years, and despite medical advances remains a major source of pain and functional limitation [12]. Compared to men, women with knee OA generally report higher pain intensity, greater perceived functional deficits and poorer quality-of-life [13].

Knee joint stability relies on a number of musculoskeletal and neuromuscular factors that counteract the stresses exerted by external loads and, therefore, muscle weakness compromises stability and balance. An inverse relationship between joint load and muscular strength implies that weak muscular support of the knee increases the load applied to the articular cartilage, which may be a risk for developing joint pathology [10]. One of the major muscle groups involved in knee joint stability is the *quadriceps femoris*. However, *quadriceps* muscle weakness in these individuals was generally attributed to disuse atrophy, which was presumed to develop because individuals minimize the use of the painful limb [10].

The evaluation of muscle strength and quality is of importance since patients with OA often show reduced muscle activity during knee extension, suggesting poorer MQ [14]. So, determining muscle atrophy and function is important in OA patients. Imaging techniques, namely ultrasound (US) [1] imaging, could be employed for assessment and follow-up of aging and muscle changes in those patients [1].

Evaluation of the abovementioned musculoskeletal structures can be diversified with respect to the imaging modality, such as Dual Energy X-ray absorptiometry (DXA), computerized tomography (CT), or Magnetic Resonance Imaging (MRI) instead of US. Despite the advantages of each modality on the musculoskeletal system, a variety of limitations can be identified, e.g., the use of ionizing radiation in longitudinal studies for CT; the cost, space, and availability of CT and MRI scanners [15, 16].

Unlike DXA, but similar to MRI, diagnostic US may be used to assess MQ via tissue characteristics [17] without applying ionizing radiation. CT can be used to evaluate the density of bony structures evaluation or to estimate body weights.

MRI is the most accurate method to evaluate joints, tendons, ligaments, and cartilage and it has great ability to display high definition images of the musculoskeletal system, becoming a potential method in clinical diagnostic and research [15, 16]. During MR imaging muscles must be followed obliquely (e.g., the *sartorius* on various cuts and sequences. However, the musculoskeletal ultrasonographer can follow this muscle from its origin to insertion during one scan. Also, if a muscle and its tendon are torn and retracted, MRI may not identify the location of the entire tendon [18]. Diagnostic US can also examine large areas, the ultrasonographer can interact with the patient to direct the examination towards the symptomatic area. It has also the advantage of being a dynamic study, in which the affected part can be imaged in real time, observing for pathologic characteristics or movement in tendon, bursa, muscles, or joints.

Diagnostic US also provides a clinically viable means to assess both muscle mass, MQ [17] and characterize normal and pathological muscle tissue [19].

Extending the findings of research reports on measurement reliability to typical clinical environments should be done with caution. The orientation of the sound transducer relative to the body surface and the compressive or shear stress on tissue through the force exerted by the examiner can alter tissue dimensions and echo-intensity (EI) [20].

US can evaluate muscle thickness (MT) and EI changes on muscle and tendons. During aging there is evidence of a decrease in MT, indicating decreased muscle mass, and increased EI indicating increased adipose content [6]. Although EI may be an important predictor of functional performance [20], it remains to be proven if screening for age-related changes in MQ may be effectively used to assess the risk of developing chronic disease and disabling conditions. In addition, the use of diagnostic ultrasound to characterize skeletal muscle have to be considered with the shortcomings related to equipment access, examiner training, limited normative datasets, and the inter-machine equivalence of EI values [17, 19].

In conclusion, diagnostic US has been used as an excellent imaging modality to evaluate the musculoskeletal system as it allows identifying the morphological change. Recent improvements in technology allow one to image tendon, muscle, joints, and even nerve with excellent resolution. The enhanced EI on ultrasonography images of skeletal muscle is believed to reflect lower MQ. Owing to the fact that EI may represent the composition of muscle with respect to infiltration of a noncontractile material, it could be a better estimator of muscle strength and functional capacity than muscle size itself [20].

Further studies are needed to understand the relationship between EI, tissue changes and functional capacity [21]. It is important to investigate age-related changes in muscle architecture, especially in elderly and very elderly individuals to facilitate the development of geriatric rehabilitation programs [7].

This chapter aims to review concepts associated with these issues, and studies that illustrate US imaging of musculotendinous structures in the elderly population during the aging process.

2 Aging and Changes in the Musculoskeletal System

Muscular changes when associated with pathology or new habits can be modified. For example, OA or physical activities can influence the muscular and tendinous morphology, and this has been the focus of current literature [22–27].

According to the 1996 US Census Bureau population projections (middle series), during the period from 1995 through 2030, the percentage of the American population ≥ 65 years and ≥ 85 years will increase by 107% and 133%, respectively [22]. Worldwide estimates predict 2 billion people will be aged over 65 years by 2050. A major current challenge is maintaining mobility and quality-of-life into old age. Impaired mobility is often a precursor of functional decline, disability, and loss of independence [28].

Based on these data, the physiologic changes that occur in skeletal muscle as a result of aging and the effects of exercise on aging skeletal muscle will be of increased importance. A Higher percentage of noncontractile tissue (fat and connective tissue) results in a decreased capacity to produce force. The change in tissue composition in older people suggests that their muscle mass may be reduced to a greater extent. The most apparent changes are decreases in muscle cross-sectional area (CSA) and the volume of contractile tissue within that CSA. Changes also occur in the function of muscle fibers, in motor unit (MU) firing characteristics, and in the aerobic capacity of skeletal muscle [22]. Maintaining good mobility in middle and old age is dependent on multiple components in muscles, bones, tendons, ligaments, and joints, and the impairment in any tissue can result in reduced mobility that in turn causes accelerated functional decline and disability [28].

Aging changes in the musculoskeletal system contribute to the development of OA by making the joint more susceptible to some risk factors, as abnormal biomechanics, joint injury, genetics, and obesity. Age is a primary risk factor for developing OA, likely due to changes in cells and tissues that make the joint more susceptible to damage and less able to maintain homeostasis [23]. Aging leads to a reduced regenerative capacity, for example, shown by the reduced levels of stem cells in the connective tissue in the elderly. The elderly enters a catabolic state, where a presenescent cell state leads to subsequent loss of connective tissue homeostasis. On a molecular level, cellular renewal, matrix protein modification, and the role of immunosenescence can affect the musculoskeletal system, which alters the bone and cartilage functions. This can demonstrate the role of aging in the development and progression of OA [24].

These muscular and tendinous changes associated with aging contribute to the difficulty experienced by elders to perform their daily tasks, such as rising from a chair, going downstairs or walking [25]. Tendons play a crucial role in movement by transmitting muscle forces to the skeleton. This may be of particular concern for older populations, who experience age-related changes in tendon composition that may affect the mechanical properties of the tendon which could eventually reduce the tendon's tolerance of mechanical loading [26]. Tendon mechanical properties are thought to degrade but improve with exercise. The mechanical behavior of tendon is nonlinear and viscoelastic. While strength and stiffness are predominantly used as distinguishing features for this structural tissue, viscoelastic behaviors are important at physiological strains. Viscoelasticity affects a tissue's ability to store, translate, and dissipate energy, and adapt to loading conditions over time [27].

Human aging is accompanied by a deterioration in endocrine functions resulting in decreased concentrations of circulating hormones. As hormones have a key regulatory role in the metabolism of different tissues, age-induced changes in the endocrine system have notable effects on the functions of different organs and body systems, as well as on functional capacity and health among middle-aged and older people. During normal aging, the most notable and well-characterized change in hormonal systems is the decrease in sex hormone production. Current literature suggests that low sex hormone concentrations are among the key mechanisms for sarcopenia and an age-induced decrease in muscle strength and power [29]. The observed deterioration of the musculoskeletal system in postmenopausal women is greater than in young women; however, it may not be due solely to aging since menopause itself might also be a contributing factor [25].

With the development of medical technology, the lifespan of humans has prolonged, and the very elderly population has grown in recent decades. It is important to investigate age-related changes in muscle architecture, especially in elderly and very elderly individuals to facilitate the development of geriatric rehabilitation programs [30]. Musculoskeletal US is a bedside imaging method used to assess muscle mass as well as architecture and composition [31].

3 Musculoskeletal Ultrasound: Concepts and Definitions

Diagnostic US is a practical approach to assess both muscle mass and MQ [17]. MQ is an important component of the functional profile of the elderly and MQ as well as muscle quantity, is reported to be associated with muscle strength [2]. Part of the decline in muscle strength, associated with skeletal muscle wastage, may be due to a reduced MQ [21]. In previous study, Carrão et al. (2015) found that quadriceps strength and enhanced EI are significantly correlated ($r = -0.511$; $p < 0.01$) [32].

Enhanced EI on ultrasonography images of skeletal muscle is believed to reflect MQ [9]. The anterior region of the thigh muscle and composition changes due to the presence of a high amount of fibrous tissue and intramuscular fat, expressed by the strengthening of EI of the muscle [2].

3.1 Ultrasound Imaging Parameters and Procedures

The accuracy of US scanning depends on operator's experience, and also on good knowledge of the anatomy and physiology to understand all the changes that occur on structures and tissues. Clearly, knowledge of the physics and parameter settings is important for obtaining good quality images. There are some fundamental steps that should be taken into account, e.g., transducer frequency, the depth adjusted to the structure, the transducer position.

Artifacts as anisotropy are very frequent and can easily suggest pathologic changes on tissues. Anisotropy is an artifact that occurs on musculoskeletal US, and it is associated with bad angulation of the insonating beam. The US beam must focus perpendicularly to the structure, promoting maximum echo return. Decreasing of the insonation angle will cause a decreasing EI of the tissue and a false positive.

It is important to have some concerns about the acquisition procedures of US images in order to have reliable results. Therefore, it is essential to control the hydration levels of individuals who need to have been at rest for some minutes before the US imaging. Also, it is crucial to use the same protocol and measurements intra- and inter-sessions. On the other hand, errors in handling the equipment can cause false results. Probe location and orientation or the compression of the probe on the skin are the most frequent errors that can occur on US scanning and the main ones that need to be controlled [33, 34].

3.2 Echo-Intensity (EI)

EI is the ability of the tissue to reflect the sound waves and produce an echo, and can give information about muscle tissue characterization [35]. The different reflections

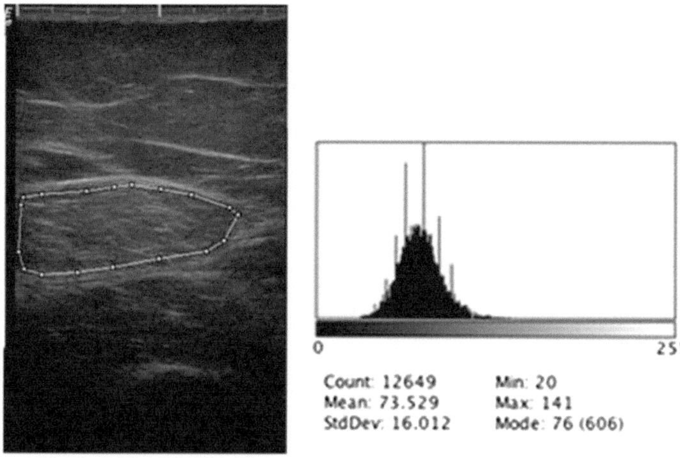

Fig. 1 Measurement of *rectus femoris* echo-intensity at midpoint between the upper pole of the patella and the iliac anterior crest of an individual with osteoarthritis of the knee [39]

of tissues will explain the different echo-intensities of each structure. Thus, the EI evaluation has an important role in muscular disorders studies [31, 35]. Morphological changes in muscles can occur due to age, sports and pathological changes [36], and several recent studies have found an association between the EI and the muscle strength, age and muscle size in middle-aged women and elderly [9, 37].

Decreased muscle mass and lipid deposition are reflected on US as an increase of muscle EI [3]. However, in tendons, the changes with aging go in the opposite direction, with EI of tendons decreasing due to the replacement of collagen by fat [2].

The EI is measured using a region-of-interest with variable size. The maximum region-of-interest is defined to include as much muscle as possible, avoiding the surrounding fascia. The mean EI of the regions may be expressed as a value between 0 (black) and 255 (white), using gray-scale analysis (Fig. 1). Alternatively, the region-of-interest can be defined as a circular or rectangular with an area chosen previously [2, 21].

3.3 *Muscle Thickness (MT)*

MT is defined as the distance between the superficial fascia of the muscle and lower boundary of its depth fascia, excluding the fascia itself [2, 21], easily determined by US (Fig. 2). MT is an important parameter related to musculoskeletal functions and has been studied in many ways and for various purposes [38]. Quantifying MT is highly reproducible. Strasser et al. (2013) found a strong correlation between quadriceps MT and maximum voluntary contraction force. They also found that

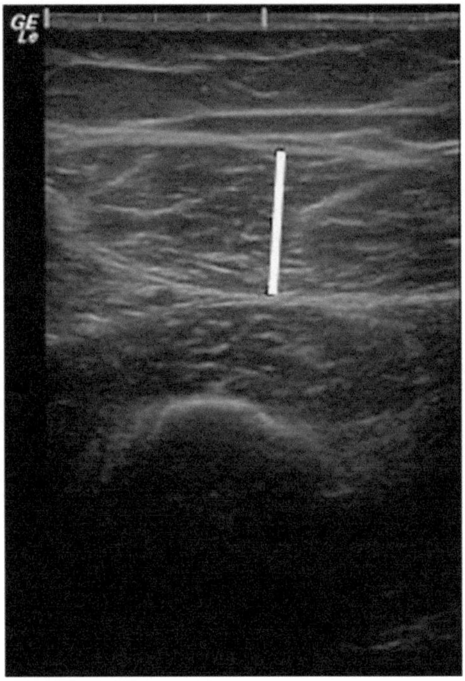

Fig. 2 Measurement of muscle thickness of *rectus femoris* at midpoint between the upper pole of the patella and the iliac anterior crest of an individual with osteoarthritis of the knee [39]

thicknesses of all muscles of quadriceps were significantly reduced in the old group compared to the young group [31].

3.4 Muscle Quality (MQ)

Importantly, changes in MQ may precede loss of muscle mass and, therefore, provide opportunities to assess MQ particularly in middle-aged adults who could benefit from interventions to improve muscle function [28]. Understanding the features of skeletal MQ changes with aging, sarcopenia, and frailty may be the key to preserving independence and reducing catastrophic events. As muscle mass and functioning decline with age, several changes occur locally within individual muscles, which affect the MQ [4].

MQ refers to the tissue's capacity to perform its several functions, including contraction, metabolism and electrical conduction. There are many relevant dimensions of MQ for muscle's capacity to perform its different functions, spanning from the broad aspect of whole muscle force production to muscle composition and morphology to the level of the sarcomere (the basic contractile element). Recent evidence suggests that the quality of muscle tissue may be more functionally relevant than its quantity [4]. MQ has been emerging as a means to elucidate and describe the intricate intramuscular changes associated with muscle performance in the context of aging and sarcopenia. While MQ has most commonly been defined in terms of muscle composition or relative strength, at the core MQ really describes the physiological functional capacity of muscle tissue [4]. MQ indices are ultimately dependent on muscle composition, architecture/morphology, and ultrastructure of the contractile *apparatus* [4]. Among future applications, one can include quantifying muscle's ability to generate force measured as strength, power, or function.

4 Discussion and Conclusions

The increase of life expectancy presents challenges since changes in muscle and tendons due to aging cause structural and functional changes in the body. Also, the increase of regular physical activity by the elderly to reduce these aging effects, raised the necessity to evaluate and follow-up, not only of the aging process but also of the results from the interventions.

The results from the aging process or from interventions in the elderly can be assessed by US imaging, which provides a noninvasive, inexpensive methodology to assess muscle morphology and estimate tissue and body composition without the use of ionizing radiation. It enables the dynamic and bilateral study of the musculoskeletal system.

Nevertheless, one should be aware of factors such as imaging site location, adequate skin contact, patient positioning, examiner training, using appropriate transducer for the specific situation, the standardization of specific assessment techniques, and the optimal use of imaging feedback may aid the wider adoption of sonography for the management of muscle disorders. Also, if we take into account the economic factors, US is less expensive and less time-dependent, in comparison with others methods such as MRI. US is a useful tool for muscular and tendinous evaluation by detecting changes in musculoskeletal disorders in the elderly after a physical activity.

Results from Kim et al. (2015) revealed loss of muscle mass in the *rectus femoris*, *tibialis anterior* and *medial gastrocnemius* in people $\geq$ 65 years old. In particular, the greatest age-related decline in muscle mass was observed for the *rectus femoris* [30]. Also Strasser et al. (2013) found that thicknesses of all muscles of the *quadriceps femoris* were significantly reduced in a group of elderly when compared to a group of young people, showing that age has is associated with a significant decrease in the thickness of the *musculus vastus lateralis* in the old group ($p = 0.0005$) [31]. When measuring echogenicity the authors found also a significant increase in the elderly, implicating that sarcopenic muscles contain more intramuscular fat and fibrous tissue than the younger subjects [31]. Rech et al. (2014) concluded that muscle EI may be an important predictor of functional performance and knee extensor power capacity in elderly, active women [20]. In a study with women with symptomatic knee OA, Carrão et al. (2015) showed an inverse correlation between EI and MT ($p < 0{,}01$) reflecting a declined MQ, as higher EI corresponds to the accumulation of noncontractile elements in the *rectus femoris* muscle [39]. Fukumoto et al. (2012) showed that physical activity can cause an increase in collagen fibers size of tendons in aged participants [2], and Santos and Amaral (2015) found a small but significant changes in EI in several muscle tendons after the program of physical activity in an elderly population [41]. Radaelli et al. (2012) and Cadore et al. (2014) argument that physical activity contributes to muscle echo-intensity reduction due to the increase in muscular vascularity and a larger amount of contractile tissue and an overall increase in muscle thickness [6, 41].

In conclusion, US imaging can be used as a gold standard method to evaluate the musculoskeletal system, as it allows identifying morphological changes in muscles and tendons, where, e.g., increased EI is believed to reflect MQ. Although EI may be an important predictor of functional performance [20], it remains to prove if screening for age-related changes in MQ may be effectively used to modify the risk of developing chronic disease and disabling conditions related to musculoskeletal health. In addition, the benefits of diagnostic US to characterize skeletal muscle have to be considered with the shortcomings of the imaging modality related to equipment access, examiner training, limited normative datasets, and the inter-machine equivalence of EI [9, 17]. Results from previous studies with elderly women showed a decline in MQ, a higher EI and lower MT [20, 31, 39]. Further studies are needed to understand the relationship between EI, tissue changes, and functional capacity.

References

1. Fechine B, Trompieri N (2012) O Processo de Envelhecimento: as principais alterações que acontecem com o idoso com o passar dos anos. Revista Científica Internacional 1(1):106–132
2. Fukumoto Y, Ikezoe T, Yamada Y et al (2012) Skeletal muscle quality assessed from echo intensity is associated with muscle strength of middle-aged and elderly persons. Eur J Appl Physiol 112(4):1519–1525. https://doi.org/10.1007/s00421-011-2099-5
3. Nishihara K, Kawai H, Hayashi H et al (2014) Frequency analysis of ultrasonic echo intensities of the skeletal muscle in elderly and young individuals. Clin Interv Aging 5 (9):1471–1478. https://doi.org/10.2147/CIA.S67820
4. Fragala MS, Kenny AM, Kuchel GA (2015) Muscle quality in aging: a multi-dimensional approach to muscle functioning with applications for treatment. Sports Med 45(5):641–658. https://doi.org/10.1007/s40279-015-0305-z
5. Bastone ADC, Jacob Filho W (2004) Effect of an exercise program on functional performance of institutionalized elderly. J Rehabil Res Dev 41(5):659–668
6. Cadore EL, Pinto RS, Bottaro M et al (2014) Strength and endurance training prescription in healthy and frail elderly. Aging Dis 5(3):183–195. https://doi.org/10.14336/AD.2014.0500183
7. Slemenda C, Heilman DK, Brandt KD et al (1998) Reduced quadriceps strength relative to body weight: a risk factor for knee osteoarthritis in women? Arthritis Rheum 41(11):1951–1959

8. Conroy MB, Kwoh CK, Krishnan E et al (2012) Muscle strength, mass and quality in older men and women with knee osteoarthritis: findings from health, aging and body composition study. Arthritis Care Res (Hoboken) 64(1):15–21. https://doi.org/10.1002/acr.20588
9. Watanabe Y, Yamada Y, Fukumoto Y et al (2013) Echo intensity obtained from ultrasonography images reflecting muscle strength in elderly men. Clin Interv Aging 8:993–998. https://doi.org/10.2147/CIA.S47263
10. Jackson BD, Wluka AE, Teichtahl AJ et al (2003) Reviewing knee osteoarthritis—a biomechanical perspective. J Sci Med Sport 7(3):347–357
11. Arendt-Nielsen L, Nie H, Laursen MB et al (2010) Sensitization in patients with painful knee osteoarthritis. Pain 149(3):573–581. https://doi.org/10.1016/j.pain.2010.04.003
12. Felson DT (2009) Developments in the clinical understanding of osteoarthritis. Arthritis Res Ther 11(1):203. https://doi.org/10.1186/ar2531
13. Tonelli SM, Rakel BA, Cooper NA et al (2011) Women with knee osteoarthritis have more pain and poorer function than men, but similar physical activity prior to total knee replacement. Biol Sex Differ pp 2–12. https://doi.org/10.1186/2042-6410-2-12
14. Ling SM, Conwit RA, Talbot L et al (2007) Electromyographic patterns suggest changes in motor unit physiology associated with early osteoarthritis of the knee. Osteoarthritis Cartilage 15(10):1134–1140. https://doi.org/10.1016/j.joca.2007.03.024
15. Dean Deyle G (2011) The role of MRI in musculoskeletal practice: a clinical perspective. J Man Manip Ther 19(3):152–161. https://doi.org/10.1179/2042618611Y.0000000009
16. Tan AL, Wakefield RJ, Conaghan PG et al (2003) Imaging of the musculoskeletal system: magnetic resonance imaging, ultrasonography and computed tomography. Best Pract Res Clin Rheumatol 17(3):513–528
17. Ismail C, Zabal J, Hernandez HJ et al (2015) Diagnostic ultrasound estimates of muscle mass and muscle quality discriminate between women with and without sarcopenia. Front Physiol 6:302. https://doi.org/10.3389/fphys.2015.00302
18. Lento P, Primack S (2008) Advances and utility of diagnostic ultrasound in musculoskeletal medicine. Curr Rev Musculoskelet Med 1(1):24–31. https://doi.org/10.1007/s12178-007-9002-3
19. Harris-Love MO, Monfaredi R, Ismail C et al (2014) Quantitative ultrasound: measurement considerations for the assessment of muscular dystrophy and sarcopenia. Front Aging Neurosci 6:172. https://doi.org/10.3389/fnagi.2014.00172
20. Rech A, Radaelli R, Goltz FR et al (2014) Echo intensity is negatively associated with functional capacity in older women. Age 36(5):9708. https://doi.org/10.1007/s11357-014-9708-2
21. Wilhelm EN, Rech A, Minozzo F et al (2014) Relationship between quadriceps femoris echo intensity, muscle power, and functional capacity of older men. Age 36(3):1113–1122. https://doi.org/10.1007/s11357-014-9625-4
22. Williams GN, Higgins MJ, Lewek MD (2002) Aging skeletal muscle: physiologic changes and the effects of training. Phys Ther 82(1):62–68
23. Loeser RF (2010) Age-related changes in the musculoskeletal system and the development of osteoarthritis. Clin Geriatr Med 26(3):371–386. https://doi.org/10.1016/j.cger.2010.03.002
24. Hügle T, Geurts J, Nüesch C et al (2012) Aging and osteoarthritis: an inevitable encounter? J Aging Res 2012:950192. https://doi.org/10.1155/2012/950192
25. Faria A, Gabriel R, Abrantes J et al (2011) Biomechanical properties of the triceps surae muscle–tendon unit in young and postmenopausal women. Clin Biomech 26(5):523–528. https://doi.org/10.1016/j.clinbiomech.2010.12.011
26. Ackermans TMA, Epro G, McCrum C et al (2016) Aging and the effects of a half marathon on Achilles tendon force–elongation relationship. Eur J Appl Physiol 116(11–12):2281–2292. https://doi.org/10.1007/s00421-016-3482-z
27. LaCroix AS, Duenwald-Kuehl SE, Brickson S et al (2013) Effect of age and exercise on the viscoelastic properties of rat tail tendon. Ann Biomed Eng 41(6):1120–1128. https://doi.org/10.1007/s10439-013-0796-4

28. McGregor RA, Comeron-Smith D, Poppitt SD (2014) It is not just muscle mass: a review of muscle quality, composition and metabolism during ageing as determinants of muscle function and mobility in later life. Longev Healthspan 3(1):9. https://doi.org/10.1186/2046-2395-3-9
29. Sipilä S, Narici M, Kjaer M et al (2013) Sex hormones and skeletal muscle weakness. Biogerontology 14:231–245. https://doi.org/10.1007/s10522-013-9425-8
30. Kim MK, Ko YJ, Lee HJ et al (2015). Ultrasound imaging for age-related differences of lower extremity muscle architecture. Phys Ther Rehabil Sci 4(1):38–43. https://doi.org/10.14474/ptrs.2015.4.1.38
31. Strasser EM, Draskovits T, Praschak M et al (2013) Association between ultrasound measurements of muscle thickness, pennation angle, echogenicity and skeletal muscle strength in the elderly. Age 35(6):2377–2388. https://doi.org/10.1007/s11357-013-9517-z
32. Carrão L et al (2015) Temperatura, força isométrica e eco-intensidade do quadricípite do joelho com osteoartrose. In Ruben RB et al (Eds). Atas do 6º Congresso Nacional de Biomecânica. Monte Real, Leiria, Portugal, pp 219–220
33. Jhu JL, Chai HM, Jan MH et al (2010) Reliability and relationship between 2 measurements of *transversus abdominis* dimension taken during an abdominal drawing-in maneuver using a novel approach of ultrasound imaging. J Orthop Sport Phys Ther 40(12):826–832. https://doi.org/10.2519/jospt.2010.3000
34. Lima K, da Matta T, de Oliveira L (2012) Reliability of the *rectus femoris* muscle cross-sectional area measurements by ultrasonography. Clin Physiol Funct Imaging 32 (3):221–226. https://doi.org/10.1111/j.1475-097X.2011.01115.x
35. Young H et al Measurement of intramuscular fat by muscle echo intensity. Muscle Nerve 52 (6):963–971. https://doi.org/10.1002/mus.24656
36. Nielsen PK, Jensen BR, Darvann T et al (2006) Quantitative ultrasound tissue characterization in shoulder and thigh muscles-a new approach. BMC Musculoskelet Disord 7(2):1–11. https://doi.org/10.1186/1471-2474-7-2
37. Pillen S, van Alfen N (2011) Skeletal muscle ultrasound. Neurol Res 33(10):1016–1024. https://doi.org/10.1179/1743132811Y.0000000010
38. Li X, Li H, Li J et al (2014) Real-Time estimation of tibialis anterior muscle thickness from dysfunctional lower limbs using sonography. HIS LNCS 8423:63–71
39. Carrão et al (2016) Relationship between quadriceps echo intensity, muscle thickness and pain intensity in aged women with knee Osteoarthritis. In BioMedWomen—international conference on clinical and bioengineering for women's Health, pp 139–142. ISBN: 978-1-317-28957-9
40. Santos R, Amaral A (2016) Ultrasound evaluation of musculotendinous structures in elderly population submitted to a physical activity program. In BioMedWomen - international conference on clinical and bioengineering for women's health, p. 135–137. ISBN: 978-1-317-28957-9
41. Radaelli R, Bottaro M, Wilhelm EN et al (2012) Time course of strength and echo intensity recovery after re-sistance exercise in women. J Strength Cond Res 26(9):2577–2584. https://doi.org/10.1519/JSC.0b013e31823dae96

Part II
Bridging Clinical and Biomechanical Aspects of Breast Surgery

Diffusion-Weighted Breast Imaging: Beyond Morphology

Luísa Nogueira, Rita G. Nunes, Sofia Brandão and Isabel Ramos

Abstract Diffusion-weighted imaging (DWI) is a magnetic resonance imaging (MRI) technique that explores the molecular diffusivity of water in biological tissues to probe its microstructure. Its application to the study of breast lesions has been shown to improve their detection, characterization, and the diagnostic accuracy of breast lesions using MRI. In this chapter, the biophysical basis of diffusion is presented, including the model currently used for DWI in the clinical setting; the concept of apparent diffusion coefficient (ADC) is introduced. A theoretical framework of DWI in healthy conditions and in tissues affected by pathological processes is presented, followed by a literature review on the application of DWI to breast imaging. As the technique has only recently been used in breast imaging studies, controversial issues regarding its application have arisen, namely related to its technical challenges. Therefore, we detail the main technical issues associated with the implementation of DWI in the clinical setting and present potential approaches for obtaining good-quality images. Finally, we identify relevant future research needs involving hardware and software optimization as well as clinical issues which need to be addressed to improve breast lesion diagnosis.

L. Nogueira (✉)
Department of Radiology, School of Health, Polytechnic of Porto (ESS-P.Porto), Porto, Portugal
e-mail: mlpnogueira@gmail.com; mlpnogueira@med.up.pt; lpn@ess.ipp.pt

R. G. Nunes
LARSyS, Department of Bioengineering, Institute for Systems and Robotics, Instituto Superior Técnico, Universidade de Lisboa, Av. Rovisco Pais 1, 1049-001 Lisboa, Portugal

R. G. Nunes
Faculty of Sciences, Institute of Biophysics and Biomedical Engineering (IBEB), University of Lisbon, Lisbon, Portugal

S. Brandão · I. Ramos
Department of Radiology, Centro Hospitalar de São João—EPE, Faculty of Medicine, University of Porto, Porto, Portugal

S. Brandão
LAETA, INEGI, Faculty of Engineering, University of Porto, Porto, Portugal

S. Brandão et al. (eds.), *Women's Health and Biomechanics*,
Lecture Notes in Computational Vision and Biomechanics 29,
https://doi.org/10.1007/978-3-319-71574-2_4

Keywords Breast imaging · Dynamic contrast-enhancement
Diffusion-weighted imaging · Apparent diffusion coefficient

Highlights

- DWI is useful in breast lesion detection and characterization
- It provides complementary information to conventional breast MRI and indirect information regarding cellular microstructure
- Different diffusion models are available to characterize signal decay
- Solutions for technical challenges associated with DWI implementation are still being actively researched

1 Introduction

1.1 Breast Imaging Modalities

Breast imaging remains an area of intensive research, and efforts to improve lesion detection and characterization data are still ongoing.

Several radiological imaging modalities are available for studying the breast. At present, mammography is the gold standard method used for screening and diagnosis of breast disease. As a screening method, it is used with the aim of detecting lesions at an early stage, preferably when women are still asymptomatic. Ultrasonography is used as an adjunct technique to mammography and breast Magnetic Resonance Imaging (MRI), and its role in the characterization of breast disease is well-established, especially in the evaluation of dense breasts; this technique aids to distinguish solid from cystic lesions and is used to guide interventional procedures. In the last years, breast MRI has taken a growing and significant role in breast imaging, despite its moderate specificity [1]. Clinical indications include the screening of high-risk patient groups and the characterization of breast disease.

The major goal of all imaging modalities when studying breast disease is to distinguish benign from malignant lesions. Although the conventional imaging methods described above provide complementary information for lesion detection and characterization, they can only be used for assessing the morphology and vascularization of the lesions. Unfortunately, these techniques are not sensitive to microscopic alterations unless they can be inferred from evident macroscopic morphological and/or vascular changes [2], limiting their ability for early detection of breast lesions.

Considering the epidemiological projections for breast cancer [3], and the central role of imaging for monitoring patients at different stages of the disease, the development of new and advanced techniques for enabling earlier detection and characterization of breast lesions is of the utmost importance.

1.2 Functional Imaging

Within the scope of MRI, several noninvasive advanced techniques have been introduced which have allowed new insights beyond conventional MRI [4, 5]. These functional techniques are perfusion, diffusion, and spectroscopy, and provide information about angiogenesis, cellularity and on the metabolic composition of the lesions, respectively. Having access to this physiological, structural, and metabolic information one can increase MRI's specificity in breast lesion characterization [4, 5].

In the clinical practice, functional imaging is used to evaluate subtle alterations occurring at an early stage, and preceding evident changes occurring from pathological processes. Recently, the inclusion of diffusion-weighted imaging (DWI) in breast MRI imaging protocols has been gaining increased importance [6]. DWI adds unique information about pathological processes and complements the information extracted from conventional breast MRI sequences.

2 Diffusion-Weighted Imaging

2.1 Biophysical Basis

DWI is a noninvasive MRI technique associated to the random thermal motion of water molecules. Through their displacement, water molecules can be used to indirectly probe characteristics of the cellular environment, such as cell density, membrane integrity, viscosity, and tissue microstructure [7]. In order to make MRI images sensitive to these local tissue characteristics, they must be sensitized to the microscopic displacement of water.

MRI relies on the excitation of nuclear spins, such as those associated to protons in water molecules, and their return to equilibrium (relaxation). When spins are subjected to a magnetic field, they precess with a frequency which is directly proportional to the intensity of this field. Excitation is achieved by applying a radiofrequency field with the matching resonance frequency. By applying a gradient field, the intensity of the magnetic field becomes spatially varying, and the resonance frequency is dependent on the spatial position of each spin.

If a gradient field was applied following MR excitation, and the nuclei were completely static, the resulting dephasing could be fully compensated for by applying a gradient with the same duration and opposite polarity. However, due to the diffusional motion of water molecules, spin rephasing is never complete, leading to signal loss. This effect can be enhanced by applying large diffusion-sensitizing gradients; signal attenuation can be modeled using a mono-exponential equation, Eq. (1):

$$S(b) = S_0 \exp(-bD),$$

where S_0 is the signal intensity obtained without diffusion gradients, S(b) is the signal intensity when diffusion gradients are used, D is the diffusion coefficient (measured in mm^2/s) and b describes the level of diffusion sensitization (measured in s/mm^2). The parameter b depends on the gradient strength (G), the time interval between the start of the two diffusion gradients (also called diffusion time $-\Delta$) and the gradient length (δ), Eq. (2):

$$b = (\gamma G\delta)^2(\Delta - \delta/3),$$

where γ is the gyromagnetic ratio characteristic of hydrogen nuclei.

To estimate the diffusion coefficient D, the same sequence can be used to generate two types of images: one with no diffusion gradients ($b = 0$ smm^{-2}), corresponding to S_0 in Eq. (1) and inherently T2-weighted; and a set of images acquired with the diffusion gradients turned on with varying amplitudes, corresponding to different b-values (diffusion-weighted images). By modeling the signal attenuation over the range of acquired b-values, D can be estimated.

When visually inspecting a DW image, one would expect that regions where water displacement is slowest to display the highest signal intensity, due to lower attenuation of the MR signal. However, as can be seen from Eq. (1), the measured signal depends not only on the diffusion attenuation, but also on the S_0 signal available. The implication is that changes in S(b) signal can arise due to changes in either or both of these factors.

S_0 depends on the proton density, T1 and T2 tissue relaxation times, as well as on sequence parameters such as the repetition (TR) and echo times (TE). Tissue changes leading to a longer T2 constant result in an increased signal in the DW image—T2 shine-through—which can erroneously be interpreted as being due to restrictions in water diffusion thus hampering evaluation of breast lesions using only visual image analyses. To overcome this artifact in the clinical practice, quantitative evaluation of diffusion is essential.

In the context of the application of DWI to biological tissues, due to the difference in scale lengths, very complex structures are captured by each individual voxel. Also, both diffusion and perfusion effects are likely to be present, which led Le Bihan et al. [8] to suggest that the term "diffusion" should be replaced by "apparent diffusion coefficient" or ADC.

In the clinical setting, often only two acquisitions with different b-values are collected to estimate ADC so as to shorten the total acquisition time. Under those circumstances, the simplest way to estimate ADC is to consider the logarithm of Eq. (1) to describe the signal attenuation between the two b-values, Eq. (3):

$$ADC = \ln[S_1/S_2]/[b_1 - b_2]$$

where b_1 is the lower b-value, b_2 the higher b-value, S_1 is the signal intensity at b_1 and S_2 is the signal intensity at b_2.

2.2 DWI in Healthy and Disease Conditions

In the human body, molecular water diffusion occurs within microenvironments having different structures. DWI is a sensitive technique, able to detect small variations in diffusion within normal tissues or in tissues altered by pathological processes. DWI is used to identify areas presenting restriction to diffusion, and through the analysis of this type of images, it is possible to obtain indirect information about tissue properties, namely its cellular organization and density, its microstructure and microcirculation.

Several compartments may be identified within biological tissues, including extracellular, intracellular, and vascular spaces. Water diffusion is restricted in the latter two due to the presence of cell membranes and, when not in the vicinity of membranes, it is hindered by cellular structures. Diffusivity depends on the integrity and permeability of the cell membrane, on the cellular viscosity and also on the blood circulation within microscopic blood vessels (microperfusion). This capillary network may, in turn, vary between tissues regarding its orientation distribution, the relative volume occupied and the blood velocity [9, 10].

The level of restriction to diffusion will depend on the cellular size and membrane permeability; that is the type of cells in each tissue governs the ADC. Therefore, based on ADC measurements, it should be possible to distinguish between tissue types, as differences at cellular level determine changes in water diffusivity.

In biological tissues, pathological processes can affect the normal configuration of intra- and extracellular spaces, changing their physical proprieties. Alterations that occur frequently are cellular dysregulation and proliferation, which cause changes in cellularity and tissue architecture, affecting water diffusivity [11].

The degree of restriction to molecular diffusivity is inversely related to tissue cellularity and integrity of cell membranes. A tissue with increased cellularity will be more compact, presenting increased restriction to water movement [12]. On the contrary, structures with low cellularity display increased diffusivity as a result of the larger extracellular space available [9].

Another relevant parameter for describing diffusion in extracellular spaces is tortuosity; its increase means that for the same effective displacement length to be observed, water molecules will have to travel a larger distance around the cells [13]. Often, increased cellularity leads to increased tortuosity in biological tissues.

2.3 Alternative Models to Describe the DW Signal Decay

As in other organs of the human body, normal tissues within the breast may be subject to alterations, cancer being the most important pathological process. Microscopic alterations at a cellular level precede macroscopic changes, undetected by conventional imaging techniques, namely mammography, ultrasonography, and conventional breast MRI. Some authors propose that the growth of breast lesions happens randomly, without structure or organized architecture [9, 11].

When applying breast DWI in the clinical setting, the diffusion displacement of water molecules is assumed to follow a Gaussian distribution. This is equivalent to assuming that diffusion is free within each voxel, and that fitting a mono-exponential model to describe the signal decay is a valid approach.

However, if we consider that diffusion within biological tissues is actually restricted, that water displacement is conditioned by the structure of lesions and tissue organization, and that breast lesions are most often heterogeneous, it is no longer possible to assume that the diffusion water displacement will follow a Gaussian distribution. To address this issue, different DWI models have been proposed to fit the signal decay and more accurately characterize breast lesions.

The first alternative model that considered non-Gaussian diffusion was introduced by Le Bihan et al. [8]. A bi-exponential model including diffusion and microperfusion (intravoxel incoherent motion-IVIM) was proposed, explaining the deviation from the mono-exponential decay at the low b-value regime due to the presence of microcirculation within randomly oriented capillary vessels.

An extension of the standard mono-exponential decay, which still relies on the assumption of free diffusion, while allowing to fully characterize the directionality of water diffusion in anisotropic tissues is the Diffusion Tensor Imaging (DTI) model. Previous results show that normal breast tissue presents larger diffusivity along the anterior to posterior direction, while pathological breast tissues display no preferential directionality, which has been interpreted as implying that breast lesions have a random growth [14].

Recently, the Diffusion Kurtosis Imaging (DKI) model has been applied to the study of breast lesions. This model is an extension of the standard mono-exponential decay, and includes an extra term, which depends on the measure of kurtosis excess; this parameter quantifies the deviation of water diffusion displacement distribution from the Gaussian shape. The kurtosis parameter has been used to evaluate the diffusion restriction resulting from the microstructural complexity of normal tissue, associated namely to the presence of barriers and of different compartments, and also to distinguish between lesion types based on differences in their microstructure [15].

3 Breast Diffusion-Weighted Imaging

3.1 Past and Current Research in Breast DWI

At present, the application of DWI for studying the breast is an area of intensive research.

The ability of DWI, conventional breast MRI and mammography for lesion detection has been studied by many researchers. Previous studies reported 6–37.5%

of lesions as not being visible in DW images [16–18]. Small lesion size and the limited spatial resolution offered by DW images are concurrent factors that impair lesion detection and localization on ADC maps [19]. Conversely, other authors reported that the combination of DWI with short-tau inversion-recovery fat suppression (STIR) presents similar sensitivity to dynamic contrast-enhancement (DCE) [20]. However, not all the lesions identified in DCE are visible on DWI [21].

DWI has been explored to evaluate occult breast lesions. Partridge et al. [22] investigated 118 lesions initially detected in breast MRI, but undetected in mammography and clinical examination. The authors found that the combination of the visual inspection of DW images and ADC enabled lesion detection and discrimination, suggesting the potential role of DWI in breast screening.

The use of DWI for detecting and characterizing axillary lymph nodes has also been studied [23]. The authors suggested that DWI is able to differentiate metastatic lymph nodes from normal lymph nodes, due to the higher signal intensity and lower ADC values found in metastatic nodes.

The diagnostic performance of DWI in lesion differentiation and characterization has been the target of many investigations. Typically, malignant lesions present lower ADC values than benign and normal tissues with significant differences between tissues types [24].

However, comparing results from different studies evidences a large variability in ADC values measured for different tissue types and some overlap in ADC estimates. The implication is that setting a cut-off ADC value for lesion classification will inevitably result in false positive and negative cases [25]. The underlying factors associated with false positive cases are the presence of edema in intra- and extracellular spaces, the existence of high viscosity areas such as those found in abscess and hematomas, of areas with high fibrosis, inflammatory changes, and of areas where blood degradation occurs. Benign lesions frequently described in the literature as false positive cases are: atypical ductal hyperplasia, papilloma, fibrocystic changes, and some fibroadenomas.

The existence of false negative results is also a cornerstone in breast DWI. It is generally accepted that malignant lesions present lower ADC values due to increased cellularity. However, some malignant lesions, especially invasive carcinomas, may present increased ADC values instead. These features are associated with low cellularity, necrotic areas, and the presence of mucinous content, increasing the ADC values. For this reason, some authors advise to jointly evaluate DWI with morphological and post-contrast images to prevent interpretation mistakes [11, 26].

The diagnostic performance of DWI depends on the ADC cut-off value selected to separate benign from malignant lesions. In the literature, ADC cut-off values range from 1.1 to 1.6×10^{-3} mm^2/s; this wide variability is related to differences in the MRI equipment and DWI sequence used, and to the heterogeneity in breast lesions included in the studies, namely, between lesion type and within the same histological type. These situations hamper the identification of a reliable ADC cut-off value to be shared between different clinical practitioners.

Non-mass lesions are not structurally formed by a single type of compact tissue, spreading instead into normal tissues; as ADC values are inversely proportional to cellular density, the restriction is lower and the ADC values are higher than in mass lesions. Non-mass lesions with these characteristics are ductal carcinoma in situ, lobular carcinoma in situ, atypical ductal hyperplasia and fibrocystic changes [26]. Classifying non-mass lesions based on ADC measurements has been reported as being particularly challenging. Many authors concluded that a higher overlap in ADC is observed for non-mass lesions resulting in decreased diagnostic accuracy [27, 28].

Different research groups have evaluated the diagnostic value of a protocol combining DCE and DWI in women with suspicious breast lesions and concluded that the combination of both examinations increased the diagnostic performance of breast MRI [29, 30].

Yili et al. [31] evaluated ADC values in malignant lesions and surrounding tissues. They found significant differences between the central region and peri-tumoral tissues, with ADC values progressively increasing from the central portion to the outermost part of the lesion. Also, they found significant differences between two neighboring layers of the lesions. The potential ability to evaluate different regions within the same lesion could be useful to determine the excision region, reducing tumor recurrence.

The relation between cellularity and ADC values has been investigated with controversial results. Some authors found an inverse relation between cellularity and ADC values [25, 32]. Contrarily, Yoshikawa et al. [33] did not find a relation with cellularity but an inverse relation could be found between ADC values and lesion histological type.

The potential use of ADC values for differentiating the histological grade of lesions has also been studied. Recently, Iima et al. [34] reported significant differences between ductal carcinoma in situ (DCIS) classified with intermediate or high grades from low-grade DCIS. They concluded that DWI shows a high specificity to identify low-grade DCIS and that the relationship between ADC values and histological grade can be used to differentiate between high and low-grade lesions.

As some studies evidence a relationship between ADC and cellularity, DWI has also been applied in the early evaluation of neoadjuvant chemotherapy and to predict treatment response. At present, this is an area of particular interest in oncology due to the possibility to select individualized therapies according to breast tumor characteristics [35].

Given the growing interest in applying DWI in the clinical practice, a recent study by Pinker et al. [36] developed a combined DCE and DWI reading based on the BIRADS-MRI lexicon. This integrated system has been shown to provide an increased specificity and accuracy in breast MRI.

3.2 Extracting Information from Breast DWI Data

It is possible to extract both qualitative and quantitative information from breast DWI images. Qualitative or semi-quantitative information can be used for lesion detection and is based on the visual analyses of DW images and ADC maps. On the native DW images, lesion detection is achieved by comparing the signal intensity of the normal tissue against that of the lesions, along the images acquired with different b-values. In contrast, in the ADC map, breast lesions are presented as areas of hypointensity, since the ADC map will most often display low diffusion coefficients in the case of malignant lesions. Figure 1 illustrates a case of a 59-year-old woman with an invasive ductal carcinoma in the left breast and includes different types of MRI images.

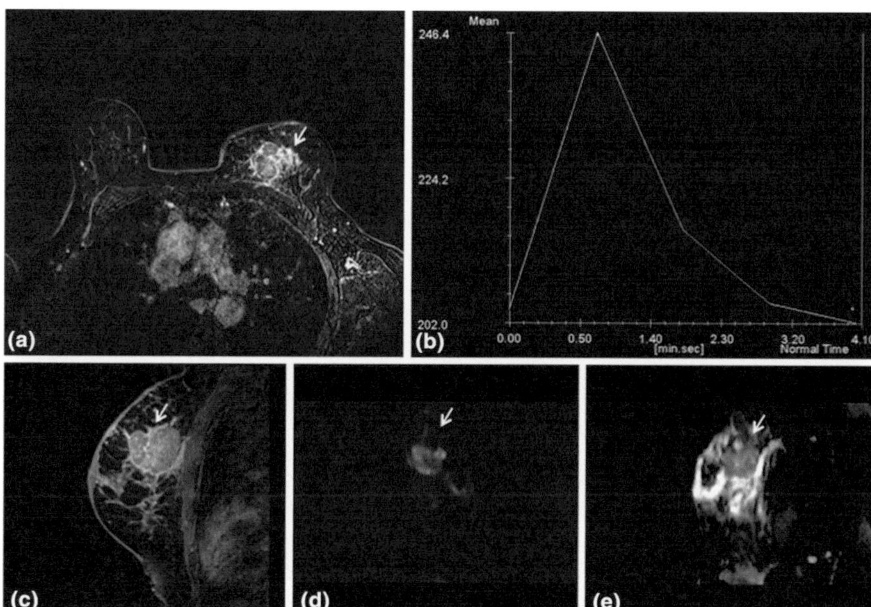

Fig. 1 A 59-year-old woman with a suspicious malignant lesion in the left breast (see arrows): **a** axial dynamic contrast-enhanced acquisition, showing a heterogeneous and irregular mass, which displays a rapid wash-in and wash-out of gadolinium (**b**). This irregular mass (**c**) shows a hyperintense signal in the sagittal diffusion-weighted image (b1000 s/mm^2) (**d**) and is hypointense on the ADC map, with a mean ADC value of 0.76×10^{-3} mm^2/s. Histological results identified the lesion as an invasive ductal carcinoma grade III with a high grade in situ component

Due to breast lesion heterogeneity, variable signal behavior can be observed on the DW images. As the b-value is increased, complex structures such as malignant lesions with high cellularity retain high signal intensity when compared to less compact or simpler structures, such as benign and cystic lesions. A steeper signal attenuation with increasing b-value is observed in lesions with low cellularity, which can condition lesion detection on images acquired with high b-values.

Quantitative information extracted from DWI studies has also been used for lesion characterization. This quantitative information is based on ADC values, measured from ADC maps. In the clinical setting, the methodology frequently used for ADC quantification is to draw regions-of-interest (ROI) on the DW images. Normally a DW image acquired with a low b (b0 s/mm^2 or b50 s/mm^2) is used for ROI demarcation, which is then copied to the other DW images and propagated to the ADC map. During ROI demarcation, areas of normal glandular tissue, necrotic areas and cystic areas and regions with high T2 signal within the lesion must be avoided, as its inclusion will bias the ADC estimates [37].

Lesion discrimination is based on the ADC values. Usually, as malignant lesions depict a more compact environment than benign ones, in these lesions ADC values are usually lower representing increased restriction [11], as in the case of Fig. 1. However, some malignant lesions do present intermediate signal intensity in DW images acquired with high b-values, and lower ADC values. Figure 2 demonstrates a case of multicentric neoplasia.

To avoid classification errors, the interpretation of DW images and the corresponding ADC map must be assessed together with the so-called "conventional" images of the breast MRI protocol.

3.3 Technical Challenges in Breast DWI

There is still no consensus between research groups regarding the optimal technical choices when implementing breast DWI in the clinical practice. These technical aspects include the basic DWI sequences and parameters that should be used,

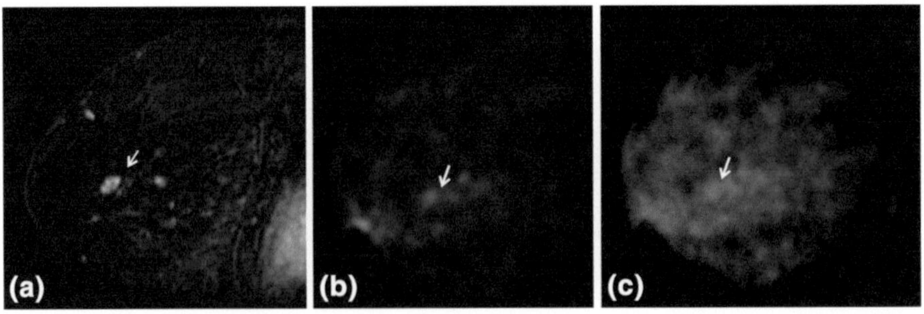

Fig. 2 A 56-year-old woman with an irregular mass of 18 mm in the left breast seen on a reconstructed dynamic sagittal (**a**) (arrow). Diffusion-weighted image at b1000 s/mm^2 (**b**) and the respective ADC map (**c**). Mean ADC value was 1.36×10^{-3} mm^2/s. Histological result: invasive carcinoma not otherwise specified grade III with intermediate grade in situ component

namely the b-values, the strategies adopted to optimize image quality, such as the fat suppression method to apply, and the methodology adopted to quantify ADC. The acquisition technique most frequently used in clinical practice is the single-shot spin-echo echo-planar imaging (SS-SE-EPI) sequence, which allows obtaining a high signal-to-noise ratio (SNR) and a very fast rate of data acquisition, minimizing movement artifacts. This is particularly important when diffusion sensitization is applied as it inherently increases sensitivity to patient motion.

Despite these advantages, the long readout of SS-SE-EPI leads to geometric distortions in the presence of static field inhomogeneities, to increased sensitivity to magnetic susceptibility differences, and chemical shift artifacts. The latter are exacerbated in DWI due to the lower ADC of fat compared to other tissues, resulting in higher signal intensity in DWI images. These issues are particularly important in breast DWI; when placing the subject within the magnet, the breasts inevitably lie away from the isocenter of the scanner, making it harder to ensure that perturbations of the static magnetic field can be corrected for. Additionally, breasts are surrounded by air and present variable amounts of fat tissue. To minimize these artifacts, shimming of the magnetic field is mandatory. Shimming relies on a dedicated set of coils to produce spatially varying magnetic fields to improve the homogeneity of the main magnetic field over the region of interest.

Although these artifacts do increase with field strength, the use of 3.0 T scanners for breast DWI can be advantageous due to the associated SNR increase. To further increase SNR, different strategies can be used, namely an increase in voxel size, in the number of excitations and TE reduction. However, the achieved TE is dependent on the chosen b-value, as longer diffusion gradients are required to increase sensitivity to diffusion. On the other hand, even for the same TE, the choice of b-value also affects the available image SNR, which decreases as the b-value increases.

The SNR is also influenced by breast density. Breasts with low density in mammography also show low SNR in DW images, especially if the suppression of fat signal is poor. To obtain a good image quality and minimize artifacts in breast DWI, a good suppression of the fat signal is crucial. There are different options available, namely through the use of chemical fat suppression techniques: frequency selective fat excitation (FatSat), water selective excitation and Dixon technique; as well as techniques based on inversion recovery: short-tau-inversion-recovery (STIR), spectral adiabatic inversion recovery (SPAIR) and spectral presaturation by inversion recovery (SPIR). It has been shown that the choice of fat suppression technique affects the contrast-to-noise ratio (CNR), SNR, and ADC estimates [38].

A poor fat suppression can affect the signal in DWI images in two manners: partial volume effects and chemical shift artifacts. The performance of the fat suppression technique determines the contribution of fat signal to the measured voxel signal intensity. As fat presents a lower ADC value compared to water, the failure of fat suppression has a more noticeable impact in ADC estimates in voxels where a large quantity of fat is present. Thus, normal glandular tissue is, in general, more affected than lesions and this effect is also more pronounced in breasts with low density. Chemical shift artifacts arise from differences in the magnetic environment of protons in water and fat molecules resulting in different resonance frequencies which in turn lead to errors in the spatial mapping of the fat signal. The long SS-SE-EPI readout can result in fat shifts of the order of a dozen voxels or even larger, and so special care is required when interpreting the DWI images as water and fat signals arising from distant sources may overlap.

The choice of the fat suppression technique to be used is dependent on the scanner's individual characteristics and performance, such as the achievable main magnetic field (B_0) homogeneity, more critical for chemical fat suppression techniques, and the baseline signal available, as inversion-recovery suppression techniques also reduce the water signal.

A controversial topic in breast DWI is the effect of the gadolinium contrast agent applied to acquire the DCE images, in the ADC value [39]. There has been some disagreement between research groups on whether the DWI acquisition should be performed before or after gadolinium injection. Rubesova et al. [19] tested the effect of gadolinium administration on the ADC by performing the DWI acquisition before and after DCE, and reported increased ADC values after injection. On the contrary, according to Yuen et al. [40], acquiring the DW images after gadolinium resulted in a 23% reduction of mean ADC values. The authors speculate that the microperfusion contribution to ADC is decreased, suggesting that estimating ADC after gadolinium injection could be more adequate for lesion differentiation.

4 Future Research

The ability to noninvasively detect microstructural changes at early stages, when they are not perceptible using conventional MR imaging techniques makes breast DWI very appealing. However, before breast DWI can be fully adopted in the clinical practice, standardization of image acquisition and data analyses protocols is required between manufacturers. To achieve this goal, it is expected that research on DWI will continue to evolve in the near future considering two main directions.

The first is related to image quality improvements that may require further developments in MRI scanners, concerning both hardware and software. DW images suffer from low SNR, which limits the achievable spatial resolution, while geometric distortions and chemical shift artifacts remain problematic. Furthermore, as SNR becomes lower, the noise distribution deviates further from a Gaussian shape; as the effect is more pronounced on high b-value DW images, it can lead to underestimation of ADC values. Software tools currently available on MR scanners do not take this effect into account, and so alternative image reconstruction or post-processing strategies are needed. Improvements in the receiver and transmit coils, as well as a wider availability of dedicated breast imaging shimming coils would also be beneficial, increasing SNR, and the homogeneity of RF and static B0 fields, which would reduce image artifacts [41].

Improving DWI image quality would also facilitate registration to DCE images. Progress in image fusion will help to take full advantage of the complementarity between sequences, and will improve breast lesion diagnosis.

Another central area to be explored in breast DWI is the development of alternative signal models, which could more directly reflect the underlying tissue microstructure. Currently, the mono-exponential model is the most frequently used for fitting the diffusion signal decay and is available on the software tools provided on clinical scanners. However, it is well known that if low b-value images are included in the fitting procedure, perfusion effects will affect the ADC values, limiting lesion differentiation and characterization. To overcome this drawback, IVIM models that allow simultaneously extracting both contributions should be provided and ideally, the choice of which model is most adequate depending on the range of sampled b-values should not require user input.

Recently, Non-Gaussian diffusion models have also been explored in breast DWI and promising results have been reported [42]. Validation of parameters arising from these models needs to be performed in the clinical setting, which will require new software to be provided in clinical MRI scanners and workstations.

Recently, an innovative approach that considers the joint effects of IVIM and non-Gaussian diffusion has been proposed by Iima and Le Bihan [43]. It is based on the differential sensitivity of the DWI signal intensity on the IVIM, ADC and non-Gaussian parameters as a function of b-value. The authors suggest considering key b-values, one in the range of "low" b-values (100–200 s/mm^2) for increased sensitivity to IVIM, and another within the "high" b-value range (1600–2400 s/mm^2) for detecting non-Gaussian diffusion. The goal is to reduce acquisition time requiring only DWI with these key b-values, while maintaining a high sensitivity to tissue structure. However, as diffusion properties are organ-specific, the best "low" and "high" key b-values to consider must be determined for breast DWI. To minimize post-processing time, and avoid noise amplification resulting from model fitting, matching the measured signal to pre-calibrated signal signatures, corresponding to benign, malignant, and normal tissue has also been suggested.

5 Conclusion

DWI is very sensitive to microstructural alterations occurring at early stages, when they are not detectable by conventional imaging modalities. DWI's proved ability for improving the specificity of breast lesion diagnosis places DWI in a singular position in the field of breast MRI. DWI is a non-invasive technique that does not require injection of contrast, which is safe, rapid and provides qualitative and quantitative information about the microstructural characteristics of lesions. Current research aims to overcome challenges so as to enable the application of DWI in the clinical setting so as to improve the diagnosis of breast cancer.

Acknowledgements RGN was funded by the Portuguese Foundation for Science and Technology FCT Investigator Program (grant number IF/00364/2013).

References

1. Peters NH, Borel Rinkes IH, Zuithoff NP et al (2008) Meta-analysis of MR Imaging in the diagnosis of breast lesions. Radiology 246(1):116–124
2. Kuhl C (2007) The current status of breast MR imaging. Part I. Choice of technique, image interpretation, diagnostic accuracy, and transfer to clinical practice. Radiology 244(2):356–378
3. Schnitt SJ et al (2014) (2014) Breast Cancer. In: Stewart BW, Wild CP (eds) World cancer report. International Agency for Research on Cancer, Lyon
4. Baltzer PA, Dietzel M (2013) Breast lesions: diagnosis by using proton MR spectroscopy at 1.5 and 3.0 T-systematic review and meta-analysis. Radiology 267(3):735–746. https://doi.org/10.1148/radiol.13121856
5. de Bazelaire C, Scémama A, Coffin A et al (2013) Perfusion studies in senology. Diagn Interv Imaging 94(12):1279–1290. https://doi.org/10.1016/j.diii.2013.06.011
6. Lo GG, Ai V, Chan JK et al (2009) Diffusion-weighed magnetic resonance imaging of breast lesions: first experiences at 3T. J Comput Assist Tomogr 33(1):63–69. https://doi.org/10.1097/RCT.0b013e318165dc6b
7. Partridge SC, McDonald ES (2013) Diffusion weighted MRI of the breast: protocol optimization, guidelines for interpretation, and potential clinical applications. Magn Reson Imaging Clin N Am 21(3):601–624. https://doi.org/10.1016/j.mric.2013.04.007
8. Le Bihan D, Breton E, Lallemand D et al (1986) MR imaging of intravoxel incoherent motions: application to diffusion and perfusion in neurologic disorders. Radiology 161(2):401–407. https://doi.org/10.1148/radiology.161.2.3763909
9. Garcia-Figueiras R, Padhani AR, Vilanova JC et al (2010) Functional imaging of tumors. Part 1. Radiologia 52(2):115–125. https://doi.org/10.1016/j.rx.2009.12.008
10. Koh DM, Padhani AR (2006) Diffusion-weighted MRI: a new functional clinical technique for tumour imaging. Br J Radiol 79(944):633–635. https://doi.org/10.1259/bjr/29739265
11. Koh DM, Collins DJ (2007) Diffusion-weighted MRI in the body: applications and challenges in oncology. AJR Am J Roentgenol 188(6):1622–1635. https://doi.org/10.2214/AJR.06.1403
12. Gore JC, Manning HC, Quarles CC et al (2011) Magnetic resonance in the era of molecular imaging of cancer. Magn Reson Imaging 29(5):587–600. https://doi.org/10.1016/j.mri.2011.02.003
13. Le Bihan D (1995) Molecular diffusion, tissue microdynamics and microstructure. NMR Biomed 8(7–8):375–386
14. Baltzer PA, Schäfer A, Dietzel M et al (2011) Diffusion tensor magnetic resonance imaging of the breast: a pilot study. Eur Radiol 21(1):1–10. https://doi.org/10.1007/s00330-010-1901-9
15. Nogueira L, Brandão S, Matos E et al (2014) Application of the diffusion kurtosis model for the study of breast lesions. Eur Radiol 24(6):1197–1203. https://doi.org/10.1007/s00330-014-3146-5
16. Park MJ, Cha ES, Kang BJ et al (2007) The role of diffusion-weighted imaging and the apparent diffusion coefficient (ADC) values for breast tumors. Korean J Radiol 8(5):390–396
17. Baltzer PA, Benndorf M, Dietzel M et al (2010) Sensitivity and specificity of unenhanced MR mammography (DWI combined with T2-weighted TSE imaging, ueMRM) for the differentiation of mass lesions. Eur Radiol 20(5):1101–1110. https://doi.org/10.1007/s00330-009-1654-5
18. Yoshikawa T, Kawamitsu H, Mitchell DG et al (2006) ADC measurement of abdominal organs and lesions using parallel imaging technique. AJR Am J Roentgenol 187(6):1521–1530. https://doi.org/10.2214/AJR.05.0778
19. Rubesova E, Grell AS, De Maertelaer V et al (2006) Quantitative diffusion imaging in breast cancer: a clinical prospective study. J Magn Reson Imaging 24(2):319–324. https://doi.org/10.1002/jmri.20643

20. Kuroki-Suzuki S, Kuroki Y, Nasu K et al (2007) Detecting breast cancer with non-contrast MR imaging: combining diffusion-weighted and STIR imaging. Magn Reson Med Sci 6 (1):21–27
21. Yabbuchi H, Matsuo Y, Sunami S et al (2011) Detection of non-palpable breast cancer in asymptomatic women by using unenhanced diffusion-weighted and T2-weighted MR imaging: comparison with mammography and dynamic contrast-enhanced MR imaging. Eur Radiol 21(1):11–17. https://doi.org/10.1007/s00330-010-1890-8
22. Partridge SC, Demartini WB, Kurland BF et al (2010) Differential diagnosis of mammographically and clinically occult breast lesions on diffusion-weighted MRI. J Magn Reson Imaging 31(3):562–570. https://doi.org/10.1002/jmri.22078
23. Kamitani T, Hatakenaka M, Yabuuchi H et al (2013) Detection of axillary node metastasis using diffusion-weighted MRI in breast cancer. Clin Imaging 37(1):56–61. https://doi.org/10.1016/j.clinimag.2012.02.014
24. Tsushima Y, Takahashi-Taketomi A, Endo K (2009) Magnetic resonance MR) differential diagnosis of breast tumors using apparent diffusion coefficient (ADC) on 1.5-T. J Magn Reson Imaging 30(2):249–255. https://doi.org/10.1002/jmri.21854
25. Guo Y, Cai YQ, Cai ZL et al (2002) Differentiation of clinically benign and malignant breast lesions using diffusion-weighted imaging. J Magn Reson Imaging 16(2):172–178
26. Brandão AC, Lehman CD, Partridge SC (2013) Breast magnetic resonance imaging: diffusion-weighted imaging. Magn Reson Imaging Clin N Am 21(2):321–336
27. Yabuuchi H, Matsuo Y, Kamitani T et al (2010) Non-mass-like enhancement on contrast-enhanced breast MR imaging: lesion characterization using combination of dynamic contrast-enhanced and diffusion-weighted MR images. Eur J Radiol 75(1):126–132. https://doi.org/10.1016/j.ejrad.2009.09.013
28. Imamura T, Isomoto I, Sueyoshi E et al (2010) Diagnostic performance of ADC for non-mass-like breast lesions on MR imaging. Magn Reson Med Sci. 9(4):217–225
29. Kul S, Cansu A, Alhan E et al (2011) Contribution of diffusion-weighted imaging to dynamic contrast-enhanced MRI in the characterization of breast tumors. AJR Am J Roentgenol 196 (1):210–217. https://doi.org/10.2214/AJR.10.4258
30. Partridge SC, Rahbar H, Murthy R et al (2011) Improved diagnostic accuracy of breast MRI through combined apparent diffusion coefficients and dynamic contrast-enhanced kinetics. Magn Reson Med 65(6):1759–1767. https://doi.org/10.1002/mrm.2276
31. Yili Z, Xiaoyan H, Hongwen D et al (2009) The value of diffusion-weighted imaging in assessing the ADC changes of tissues adjacent to breast carcinoma. BMC Cancer 9:18. https://doi.org/10.1186/1471-2407-9-18
32. Hatakenaka M, Soeda H, Yabuuchi H et al (2008) Apparent diffusion coefficients of breast tumors: clinical application. Magn Reson Med Sci 7(1):23–29
33. Yoshikawa MI, Ohsumi S, Sugata S et al (2008) Relation between cancer cellularity and apparent diffusion coefficient values using diffusion-weighted magnetic resonance imaging in breast cancer. Radiat Med 26(4):222–226. https://doi.org/10.1007/s11604-007-0218-3
34. Iima M, Le Bihan D, Okumura R et al (2011) Apparent diffusion coefficient as an MR imaging biomarker of low-risk ductal carcinoma *in situ*: a pilot study. Radiology 260(2):364–372. https://doi.org/10.1148/radiol.11101892
35. Richard R, homassin I, Chapellier M et al (2013) Diffusion-weighted MRI in pretreatment prediction of response to neoadjuvant chemotherapy in patients with breast cancer. Eur Radiol 23(9):2420–2431. https://doi.org/10.1007/s00330-013-2850-x
36. Pinker K, Bickel H, Helbich TH et al (2013) Combined contrast-enhanced magnetic resonance and diffusion-weighted imaging reading adapted to the "Breast Imaging Reporting and Data System" for multiparametric 3-T imaging of breast lesions. Eur Radiol 23(7):1791–1802. https://doi.org/10.1007/s00330-013-2771-8
37. Koh D, Takahara T, Imai Y et al (2007) Practical aspects of assessing tumors using clinical diffusion-weighted imaging in the body. Magn Reson Med Sci 6(4):211–224

38. Nogueira L, Brandão S, Nunes RG et al (2015) Breast DWI at 3T: influence of fat suppression technique on image quality and diagnostic performance. Clin Radiol 70(3):286–294. https://doi.org/10.1016/j.crad.2014.11.012
39. Nguyen VT, Rahbar H, Olson ML et al (2015) Diffusion-weighted imaging: effects of intravascular contrast agents on apparent diffusion coefficient measures of breast malignancies at 3 Tesla. J Magn Reson Imaging 42(3):788–800. https://doi.org/10.1002/jmri.24844
40. Yuen S, Yamada K, Goto M et al (2009) Microperfusion-induced elevation of ADC is suppressed after contrast in breast carcinoma. J Magn Reson Imaging 29(5):1080–1084. https://doi.org/10.1002/jmri.2
41. Lee S-K, Hofstetter L, Hancu I (2011) B0 shimming in 3T bilateral breast imaging with local shim coils. Proc Intl Soc Mag Reson Med 19:715
42. Iima M, Yano K, Kataoka M et al (2015) Quantitative non-Gaussian diffusion and intravoxel incoherent motion magnetic resonance imaging differentiation of malignant and benign breast lesions. Investig Radiol 50(4):205–211. https://doi.org/10.1097/RLI.0000000000000094
43. Iima M, Le Bihan D (2016) Clinical intravoxel incoherent motion and Diffusion MR imaging: past, present and future. Radiology 278(1):13–32. https://doi.org/10.1148/radiol.2015150244

The Effects of Mastectomy and Breast Reconstruction on Body Posture and Biomechanical Aspects

Ana Paula Ribeiro, Thalissa Maniaes and Adriana Naomi Hamamoto

Abstract The purpose of this chapter is to highlight the main changes on the biomechanical aspects focused in posture, body balance and gait of women with breast cancer after surgical treatment of mastectomy, with or without placement of the breast prosthesis. In summary, women with mastectomy showed posture changes, such as asymmetry of trunk and shoulder girdle and greater forward leaning of the trunk. Furthermore, were observed asymmetry of the pelvis, scapula and shoulder one year after the mastectomy with radiotherapy. Women's with mastectomy showed lymphedema, head rotation to the right, protrusion of the left shoulder, and trunk inclination angle smaller on the operated side, besides bilateral elevation of the scapula. However, women's post immediate breast reconstruction, with abdominal flaps, demonstrated asymmetry in the vertical alignment of the trunk with greater asymmetry between the acronyms and greater trochanters, which can change trunk balance. The weight of external breast prosthesis can also contribute to posture changes in women post mastectomy, such as asymmetry of the head due to erector spinal muscle activity imbalance in women with mastectomy. Changes in the gait parameters-velocity, cadence, time and step length also were observed with and without prosthesis. These clarifications may provide reliable information for the prescription of physical exercise, such as intervention protocols with exercise for posture and upper kinetic chain, as well as feet motion-control during gait for better postural balance after mastectomy or placement of the prosthesis.

Keywords Breast cancer · Posture · Balance · Gait · Mastectomy
Prosthetic reconstruction

A. P. Ribeiro (✉)
Department of Physical Therapy, Faculty of Medicine, University of São Paulo, São Paulo, Brazil
e-mail: anapaulafisioterapia@yahoo.com.br

A. P. Ribeiro · T. Maniaes
Department of Health Science, Post-Graduate and Physical Therapy, University Santo Amaro, São Paulo, Brazil

A. Naomi Hamamoto
Cancer Institute Hospital, Faculty of Medicine, University of São Paulo, São Paulo, Brazil

S. Brandão et al. (eds.), *Women's Health and Biomechanics*,
Lecture Notes in Computational Vision and Biomechanics 29,
https://doi.org/10.1007/978-3-319-71574-2_5

Highlights

- Surgical treatment of breast cancer with mastectomy and with or without breast prosthesis is, to both patients and health professionals, a rehabilitation process that involves improving the clinical and functional conditions of the bodily systems
- The chapter presents some adaptations of the body system of women after mastectomy and after different types of breast implant
- Clarifications on biomechanical aspects (posture balance and gait after mastectomy or breast implant) may provide reliable information for the prescription of physical exercise and rehabilitation program
- The biomechanical analysis of the bodily systems is the key to improvement functional after surgical treatment of the breast cancer

1 Introduction

Breast cancer stands for nearly 12% of all new malignant neoplastic cases and almost 25% of all cancers in women worldwide with nearly 1.7 million new cases identified each year [1]. Generally, is more prevalent in women over 40 years of age with consequent changes in the musculoskeletal system [2]. In recent years, there has been a significant increase in the prevalence of breast cancer also in younger women. Young age at diagnosis of breast cancer in women is associated with more aggressive disease and worse clinical outcomes. However, there is no consensus on the exact age that defines "young" women, and numerous investigators have used cut-offs ranging from less than 35 years up to less than 50 years of age. Given the small numbers of young women in these studies uncertainty persists about the optimal primary treatment for this risk population [3].

Due to this alarming increase in incidence, breast cancer is considered a public health problem, requiring a complex and multidisciplinary treatment approach ranging from promotion and prevention to clinical curative and therapeutic or even palliative treatments, promoting full attention to the health of these women [4, 5].

Large proportion of breast cancer patients qualifies for breast conservation therapy comprising tumor excision, sentinel node biopsy and radiation therapy. Unfortunately, there is still a range of contraindications for this procedure (i.e., multicentric disease, inflammatory breast cancer, expected unfavorable cosmetic outcome, BRCA1 and BRCA2 mutations, previous radiotherapy), which results in several patients requiring mastectomy. Those patients are potential candidates for reconstructive procedures that can be performed during primary surgery or as an independent procedure [1].

When surgical treatment by mastectomy or breast reconstruction occurs, a multidisciplinary team composed of clinical doctor, surgeon, physical therapist, and psychologist are justified based on the large biopsychosocial impact on women undergoing mastectomy and their family members [5].

The literature has shown that after mastectomy or breast reconstruction posture adaptations [6–10] and gait biomechanical changes [11] can occur in women with breast cancer, and that they may interfere in maintaining balance postural these women [12, 13]. The results of different studies have reaffirmed that posture changes following mastectomy and breast reconstruction usually resulting in functional upper kinetic chain adjustments [14–16].

The process after mastectomy involves the consideration of restorative options, including the use of an external prosthesis. The prosthesis has been considered very important for body image, psychosocial well-being and functional aspects that greatly improve quality of life in women who have undergone a mastectomy [16–18].

Previous studies about biomechanical motion analysis focused on posture assessment [10] and the analysis of muscle strength [19], but failed to consider functional aspects. Taking into consideration the changes in body posture after mastectomy or after breast reconstruction [10] and the influence of spine kinetic disturbances on the dynamic parameters of gait, it appears that the appropriate choice of prosthesis may also influence whole-body movement dynamics [11]. However, the literature has not yet reported studies on clinical intervention programs directed at minimizing these changes in posture and gait during pregnancy.

A deep comprehension of posture, balance and gait changes in women with breast cancer is needed to improve physiotherapeutic treatment. These clarifications may provide reliable information for the prescription of physical exercise, such as intervention protocols with exercise upper kinetic and feet motion-control during gait for better postural balance of the women's after mastectomy or placement of the prosthesis.

The purpose of this chapter is to highlight the main changes on the biomechanical aspects focusing in posture, body balance and gait of women with breast cancer after mastectomy surgical treatment with or without reconstruction with breast prosthesis.

2 Epidemiology and Pathologic Features of Breast Cancer

Breast cancer is the second most common cancer in the world with more than 1 million new cases and also second cause of death, and it is a major public health problem in developed, underdeveloped or under development countries [20, 21]. Mortality from breast cancer follows the same socioeconomic gradient as incidence [22]. Women in higher socioeconomic groups are more likely to have breast cancer recorded as their cause of death than those in lower socioeconomic groups. However, the survival in more deprived groups is worse at every stage of the disease [23]. Studies have shown that women from lower socioeconomic backgrounds are more likely to be diagnosed with more advanced disease [24], with differences being more pronounced in the range of 50–69 years [25] and are more likely to have a poorer prognosis than affluent women. This relates to the fact that women from deprived groups are less likely to undergo screening examinations [23].

All ages and ethnicity are affected: African, American, Hispanic and Asian women will develop breast cancer during their lifetime. Young age has been identified as a risk factor for recurrence and death from breast cancer. Some studies have suggested that young women (less than 35–40 years of age) have associated with more aggressive disease and worse clinical outcomes [3].

According to the National Cancer Institute (INCA) [26], estimates for 2012–2015 indicate that 52,680 new cases of female breast cancer will be identified, which it corresponds to the occurrence of approximately 52 cases per 100,000 women. Of these new cases in women and men, around 10% are diagnosed in the advanced stages, when the tumor has spread significantly within the breasts or have disseminated metastasis to other organs of the body. Despite advances in early detection and the understanding of the molecular bases of breast cancer biology, about 30% of patients with early-stage breast cancer have recurrent disease [21].

Breast cancer is a complex disease. Breast cancers that haven't grown into the stroma are called "noninvasive" and breast cancers that have grown into the stroma are called "invasive". Invasive cancer cells can leave the breast and form tumor in other parts of the body, process called metastization [27].

The breast cancer will be categorized into 1 of 5 stages. Treatment choices will depend on tumor size and location in the breast; if cancerous cells are found in the lymph nodes in armpit or if there is evidence of metastatic disease. The 5 stages are [27]:

- ***Stage 0***: corresponds to a very early breast cancer or pre-invasive cancer, which has not spread within or outside of breast (also called in situ or noninvasive cancer);
- ***Stage 1***: smaller than 2 cm; with no cancerous cells found in lymph nodes in the armpit, or outside the breast;
- ***Stage 2***: smaller than 2 cm; with cells found in the lymph nodes in the armpit, OR tumor size between 2 and 5 cm, with cancer cells that may or may not be found in the lymph nodes in the armpit, OR tumor larger than 5 cm, with no cancer cells found in the lymph nodes in the armpit.
- ***Stage 3***: Tumor smaller than 5 cm with cancer also in the lymph nodes that are stuck together, OR Tumor larger than 5 cm, OR cancer is attached to other parts of the breast area including the chest wall, ribs, and muscles, OR inflammatory breast cancer, rare type of cancer, the skin of the breast is red and swollen.
- ***Stage 4 or 5***: tumor metastasis spread to the body (e.g., bones, lungs, liver, or brain).

When cancer is detected early, five-year survival rates are very high. Almost all women with Stage 0 cancer will have a normal lifespan. Five-year survival rates are as high as 95% when the cancers in Stage 1 are smaller than one centimeter. Even when a cancer falls into a Stage 2 category, five-year survival rates are close to 70% [28].

Some women are at higher risk for the spread and relapse of breast cancer. The risk factors for recurrence are complex, they are not absolute forecasts of what the future will be. The factors are:—***Tumor size***: the larger your tumor, the higher the risk;—***Lymph nodes***: the more lymph nodes in the armpit ipsilateral to the breast, the higher your risk;—***Cell studies***: new tests can measure the growth rate and aggressiveness of the tumor cells. The cancer cells that show the most rapid growth are linked to higher risk for the return of cancer [28].

3 Types of Breast Cancer, Diagnosis, and Treatment

Despite advances in early detection and the understanding of the molecular bases of breast cancer biology, about 30% of patients with early-stage breast cancer have recurrent disease. Two types exist of breast cancer [27]:

- Ductal carcinoma in situ (DCIS) is noninvasive, which means it is limited to the breast ducts. However, there are several types of DCIS. If it is not removed, some types may in time change and develop into its invasive form.
- Lobular carcinoma in situ (LCIS) is a noninvasive growth limited to the lobules. It is not cancer, only a warning sign of increased risk of developing cancer, according to the National Cancer Institute [27]. Women with LCIS have about a 1% risk of developing invasive lobular carcinoma equally in either breast per year. At 20 years, this risk is about 18%.

According to the literature, the first step is to establish a medical history, including any health events that can run in families or due to a hereditary BRCA 1 and BRCA2-associated increases risk to breast cancer. Second, the physical exam is a study of body for signs of disease, searching for (ab) normal size, softness, or pain, and also evaluation of possible lymph nodes. Third, diagnostic mammogram and/or ultrasound are current diagnostic screening tools in most countries to confirm and characterize the suspicious lesion, and to search for bilateral disease. Fourth, a biopsy is then used to remove part of the mass for testing abnormal cells [28].

After performing these tests for an accurate and reliable diagnosis, the treatments may be directed to lumpectomy and/or radiation therapy or mastectomy. Literature shows that both general options provide the same long-term survival rates. However, neither gives you full guarantee that cancer will not relapse [28].

The lumpectomy includes excising the lesion and normal breast tissue around, as well as existing lymph nodes under the arm. This procedure aims at removing the cancer while leaving the breast much the same as it did before your surgery. Women who choose a lumpectomy almost always have radiation therapy as well to decrease the risk of cancer in the remaining breast tissue [28, 29].

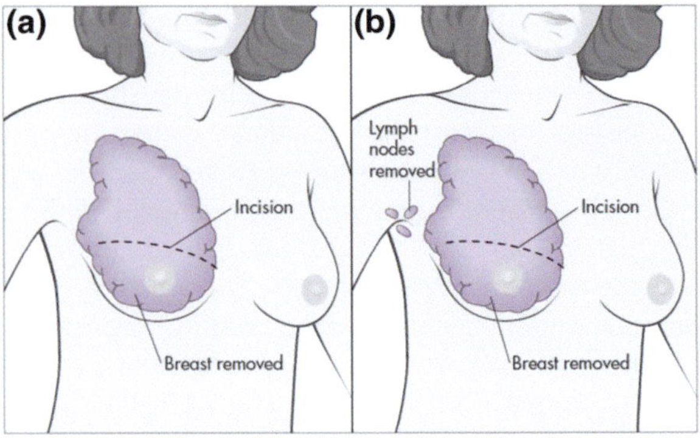

Fig. 1 Illustration **a**: total mastectomy and **b**: modified radical mastectomy Illustration copyright © 2016, Developed by the California Department of Health Services, A Woman's guide to breast cancer diagnosis and treatment. All rights reserved. California Department of Health Services

Mastectomy includes removing the breast, and it is needed when: the cancer is found in numerous areas; the breast is small or shaped so that removal of the entire cancer will leave little breast tissue or a deformed breast or the woman does not want to have radiation therapy [27]. Total mastectomy includes removing as much breast tissue as possible, the nipple, and some of the overlying skin. The lymph nodes in the armpit are not removed. Modified radical mastectomy is a surgery removes as much breast tissue as possible, the nipple, some of the overlying skin, and some lymph nodes in the armpit, as shown Fig. 1.

Surgical treatment of breast cancer has been marked by a constant evolution since the Halsted radical mastectomy described in the late nineteenth century. The current procedure is the standard Madden radical mastectomy, a breast surgery that involves the ablation of tissue with the axillary lymphatic preserving both pectoral muscles [30]. A good medical outcome is often reached at an important psychological price associated with post mastectomy depression and lower self-image. Therefore, a significant improvement of patient's quality of life can be obtained thanks to immediate or delayed breast reconstruction [1].

Breast reconstruction was initially created to reduce complications of mastectomy and to diminish chest wall deformities. However, it is currently known that breast reconstruction can also improve the psychosocial well-being and quality of life of patients. Its primary goal is to recreate form and symmetry by correcting the anatomic defect while preserving safety. The primary reconstructive options involve the use of an implant (usually with an expander first), the patient's own tissue (autogenous tissue reconstruction), or both. The reconstructive process can start at the time of the mastectomy (immediate reconstruction) or any time afterwards (delayed reconstruction) [31].

In 60–70% of the cases, the reconstruction can be performed with an implant inserted behind the major *pectoralis* muscle. The expanders are inflated progressively in the postoperative course thanks to a reservoir located subcutaneously to provide progressive distention of the teguments and a more natural shape after replacing by a definitive implant. The symmetry is usually obtained through contralateral plastic surgery, which allows at the same time histological check up of the glandular tissue. The nipple areolar complex is usually reconstructed in a second stage under local anesthesia, using local flaps for the nipple and a tattoo to color the areola [31, 32].

In 30% of the cases, especially after radiotherapy when a salvage mastectomy is required, a flap reconstruction is preferred. Muscle, fat, and skin from another part of the body can be moved to the chest area, where it is shaped into the form of a breast. This tissue can be taken from the lower stomach area (*rectus abdominis* muscle flap), back (*latissimus dorsi* muscle flap) or buttocks (*gluteus* muscle flap), as show Fig. 2. The autologous tissue reconstruction with the *rectus myocutaneous* flap gives excellent cosmetic results and the most natural shape for the breast, but it is a more demanding technique requiring a good experience. In some occasions, the reconstruction with the *latissimus* flap can also be autologous but usually requires the addition of prosthesis. In most cases, the reconstruction can be performed immediately. The delayed reconstruction is usually preferred when the adjuvant chemotherapy should be delivered as soon as possible after the mastectomy [31, 32].

Complications of the reconstruction procedure, such as local necrosis or infections, may lead to implant removal or revision of the flap, and can be detrimental to the patient in delaying the start of the chemotherapy. It is not recommended to reconstruct the breast immediately in case of locally advanced breast cancer. Partial breast reconstruction using plastic surgery procedures can also be performed in case of quadrantectomy in order to obtain a better cosmetic result. Local glandular flaps, as well as specific incisions according to the location of the tumor in the breast allow the reshaping of the breast even in case of large resection and, therefore, provide an opportunity to increase the number of conservative treatment indications, especially in case of in situ carcinomas [1, 32].

Generally, after surgical mastectomy or reconstruction of breast some post-operative complications can occur on the musculoskeletal system, usually associated with axillary retractable scar, pain syndrome, lymphedema, upper limb mobility limitation [33, 34]. Furthermore, mastectomy without breast reconstruction is a very effective and curative often procedure, but has a profound mutilating effect that inevitably influences the patient psychologically, often causing not only negative body image and anxiety, but also depression. After mastectomy, most women are given the option of either immediate or delayed breast reconstruction. Although these procedures are intended to provide cosmetic and psychological improvements, such as possible reduction of depression when compared to mastectomy without breast reconstruction (BR), contrasting reports have been written about these improvements [35].

Multiple studies noted a significant decrease in depression in the reconstruction group, whereas in a number of other studies no psychological improvements were found in women's with both breast reconstruction groups (immediate or delayed) during follow-up periods of 6 months to 2 years after surgery. In view of the clinical possible consequences of breast reconstruction, such as prolonged recovery time, risks of complications or the need of additional surgery, data should be available regarding the advantages of these surgical procedures including the

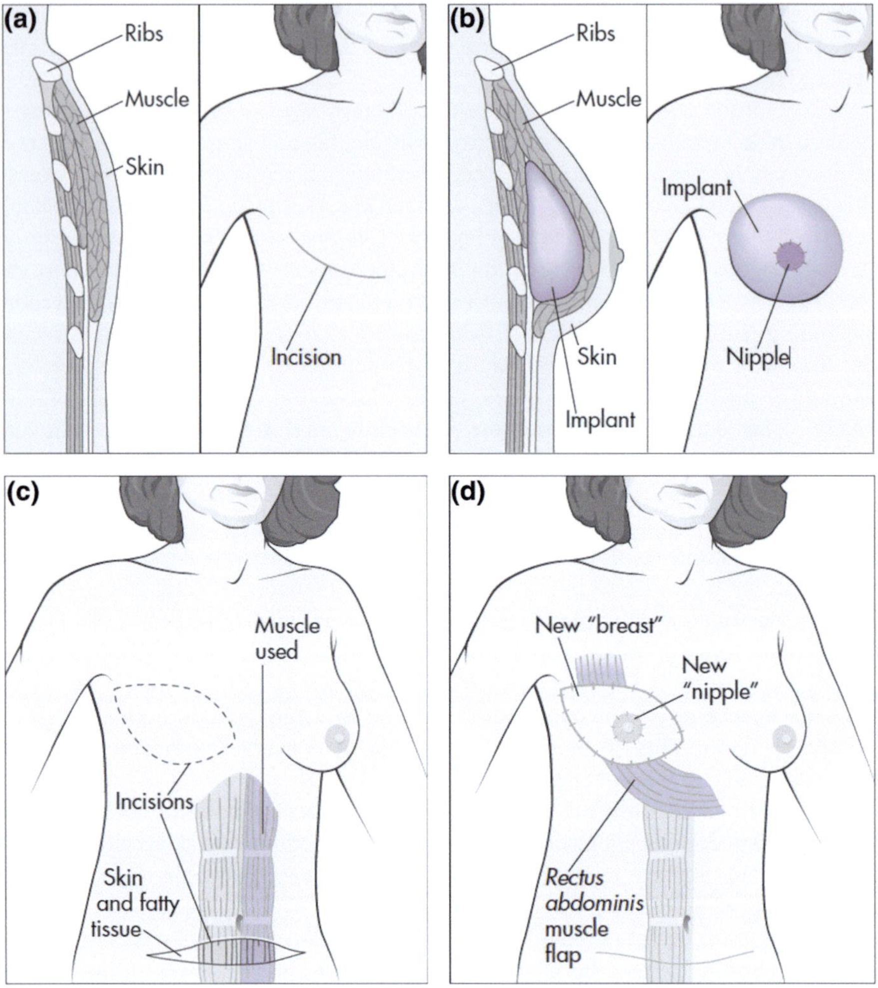

Fig. 2 Illustration **a**: total mastectomy; **b**: breast reconstruction; **c**: flap of muscle, skin and fatty tissue and **d**: flap of the rectus abdominis muscle Illustration copyright © 2016, Developed by the California Department of Health Services, A Woman's guide to breast cancer diagnosis and treatment.

influence on depression risk with physical changes important in body of women [35].

4 Changes in Body Posture After Mastectomy and Breast Reconstruction

Presently, in addition to these possible physical on upper kinetic chain and emotional changes after surgery treatment of breast cancer (mastectomy and reconstruction), some studies have focused on the clinical importance of understanding the postural adaptations and biomechanics aspects directed for body balance and gait of women with breast cancer after mastectomy and reconstruction of the breast.

A large number of women complain of increased back pain a few months to years after mastectomy. The study of Rostkowska et al. (2006) has showed a statistically significant disturbance of proper body posture in women after mastectomy [6]. The research included 85 women after mastectomy and 80 healthy women from the control group, and the patients manifested asymmetry of trunk and shoulder girdle and greater forward leaning of the trunk. Significant relationship was indicated between mastectomy and the asymmetry of position of the scapulas.

Malicka et al. (2010) evaluated the body posture of women after mastectomy and observed that 82.3% of women after treatment presented with posture failures, against 35.1% in healthy women [7]. According to authors, women having undergone mastectomy who also participated in rehabilitation are characterized by diverse posture types: kyphotic, lordotic and balanced. Women with lordotic posture achieved the highest force-speed parameter values in the trunk muscles irrespective of the muscle group examined, while the lowest values were observed in women exhibiting kyphotic posture. The function of trunk muscles in women following mastectomy depends on body posture type. In addition to postural changes resulting from mastectomy it is important for health professionals, especially physiotherapists, understand the changes of posture after the techniques different for breast reconstruction.

Currently, there are many options for breast reconstruction surgery, and the impact of the different techniques on body posture has been studied. One study demonstrated that immediate breast reconstruction with a Beker-25 prosthesis could help to preserve body posture after mastectomy in difference coronal, sagittal and transverse planes compared groups of patients after mastectomy (18 and 24 post-surgical months) [8].

Another important aspect is evidence regarding the effect of surgery on the body posture of women after breast reconstruction when using autologous tissue. A study conducted by Peres et al. (2015) [9] observed that women who underwent mastectomy alone (n = 38) compared with women who underwent immediate breast reconstruction with abdominal flaps (n = 38) showed differences in the vertical alignment of the trunk, with greater asymmetry between the acromion and greater trochanter, which can mean trunk rotation.

Głowacka et al. (2016) [10] assessed the magnitude of frontal plane postural changes in breast cancer survivor patients after a follow-up 1 years of mastectomy with breast-conserving therapy. The authors observed highly significant differences in shoulder line angle values 2 months after the mastectomy with radiotherapy, as well as highly significant differences in pelvic inclination angle, scapular angle, distance of the shoulder blades from the spine and shoulder line angle values 1 years after the mastectomy with radiotherapy.

Alterations in posture were also observed in women after mastectomy and lymphadenectomy, submitted to radiotherapy as adjuvant treatment. The women with mastectomy and lymphedema had head rotation to the right, protrusion of the left shoulder, and trunk inclination angle smaller on the operated side, besides bilateral elevation of the scapula when compared to the group with no lymphedema [36].

A recent study conducted by Hojana et al. (2016) [37] evaluated and determined whether the weight of an external breast prosthesis can contribute to posture changes in women post mastectomy. The *major* findings were: (1) whether that the surgery was done on the left or right side, the erector spinal muscle activities on the ipsi—and contralateral side were significantly different; (2) overall, for patients who had operation on the left side, their erector spinal muscle activity imbalance was smaller when compared to the imbalance among patients with right breast. This relationship exists even with all other factors were adjusted; (3) the weight of the external breast prosthesis did not influence the differences of erector spinal muscle activities on the operation side and non-operated side. Together with the postural changes, adaptive musculoskeletal changes also occur that may interfere with maintaining vertical balance.

5 Changes in Postural Balance and Gait Parameters After Mastectomy

Postural balance is defined as the perception and control of motion of the musculoskeletal system. It is influenced by three collaborative systems, namely the somatosensory, vestibular, and visual. Every new posture adopted by humans triggers neuromuscular responses necessary to maintain body balance. Among these responses, the sensory system provides information about the position of body segments relative to other segments and to the environment; the motor system is responsible for correct and proper activation of the muscles; and the central nervous system integrates the information from the sensory and the motor systems [38].

The literature shows the relevance of checking the balance and postural and functional changes in women after surgical treatment of breast cancer or radical mastectomy. These balance assessments enable a broader vision of functional rehabilitation, the purpose of which is to address the consequent physical and functional aspects of breast cancer treatments.

Only two studies focused on balance after mastectomy and breast reconstruction. The first evaluated the influence of mastectomy in postural control of women undergoing surgical treatment of breast cancer. Women in the mastectomy group showed an increase in the displacement area along the lateral (x-axis) and total displacement with eyes open (EO), as well as an increase in the x-axis and total displacement with eyes closed (EC). Therefore, the results showed that surgery for unilateral mastectomy may significantly alter postural control [12].

The second study evaluated the importance of maintaining control patterns in static equilibrium after mastectomy, and the authors found statistical differences within a significant part of the parameters measured with EC, and for all the Romberg parameters: COP (centre of pressure); path length measured in the anterior-posterior direction; the average tilt COP the maximum swing in the x-axis (related to the range lateral stability). The authors concluded that the postural balance was less dependent on the vision in women subjected to mastectomy than in the control group. Physiotherapy programs after mastectomy should include proprioceptive training, with EC to improve the equilibrium reaction quality and increase the postural stability [13].

These changes of postural balance after mastectomy can suggested adjust step during gait as a form of adaptation postural instability and result in compensatory mechanism to effective movement control in perform functional tasks. However, only one study was lead to investigate the possible gait adjustments in women after mastectomy with breast reconstruction. Hojan et al. (2013) [11] evaluated the spatiotemporal gait parameters of 40 women after mastectomy with external breast prosthesis. Significant differences were found in the gait parameters—velocity, cadence, step time (right and left) and step length—of the "younger age" group, with and without prosthesis. No significant differences were found in the women of the "older group", with and without the prosthesis. Gait parameters of the "younger age" group were closer to those of the healthy control group when they were wearing an external prosthesis, as compared with when they were not. This suggests a positive influence of breast prosthesis use on the functional status of women after mastectomy.

In summary, women with mastectomy showed posture changes, such as asymmetry of trunk and shoulder girdle and greater forward leaning of the trunk. Furthermore, asymmetry in the pelvic inclination angle, scapular angle or distance of the shoulder blades from the spine can be found 1 year after the mastectomy with radiotherapy. Women with mastectomy and lymphedema demonstrate head rotation to the right, protrusion of the left shoulder, and trunk inclination angle smaller on the operated side, besides bilateral elevation of the scapula. Mastectomy alone compared with women who underwent immediate breast reconstruction with abdominal flaps was associated with asymmetry in the vertical alignment of the trunk with greater asymmetry between the acromion and greater trochanter, which can mean trunk rotation. The weight of external breast prosthesis can also contribute to posture changes in women post mastectomy, such as asymmetry of the head due erector spinal muscle activity imbalance.

Balance postural and gait studies showed an increase in the displacement area along the lateral x-axis and total displacement with eyes open (EO) and closed (EC) in women with mastectomy and changes on the gait parameters—velocity, cadence, time and step length, with and without prosthesis.

Further research, with larger populations is still needed to further study postural balance and gait parameter in women's that perform mastectomy or BR. These clarifications may provide reliable information for the prescription of physical exercise, such as intervention protocols with exercise and feet motion-control during gait.

References

1. Szloch J, Marczyk E, Kołodziej-Rzepa M et al (2016) Impact of different type of cancer treatment on the effectiveness of breast reconstruction. Gland Surg 5(4):444–449. https://doi.org/10.21037/gs.2016.05.06
2. Bertelsen L, Mellemkjaer L, Christensen J et al (2009) Age-specific incidence of breast cancer in breast cancer survivors and their first-degree relatives. Epidemiology 20(2):175–180. https://doi.org/10.1097/EDE.0b013e318190eee6
3. Cao JQ, Olson LA, Tyldesley SK (2013) Comparison of recurrence and survival rates after breast-conserving therapy and mastectomy in young women with breast cancer. Curr Oncol 20(6):593–601. https://doi.org/10.3747/co.20.1543
4. Guerra MR, Gallo CVM, Mendonça GAS (2005) Risco de câncer no Brasil: tendências e estudos epidemiológicos mais recentes. Rev Bras de Cancerol 51(3):227–234
5. Bastos J, Peleteiro B, Gouveia J et al (2010) The state of the art of cancer control in 30 European countries in 2008. Int J Cancer 126(11):2700–2715. https://doi.org/10.1002/ijc.24963
6. Rostkowska E, Bąk M, Samborski W (2006) Body posture in women after mastectomy and its changes as a result of rehabilitation. Adv Med Sci 51:287–297
7. Malicka I, Hanuszkiewicz J, Stefanska M et al (2010) Relation between trunk muscle activity and posture type in women following treatment for breast cancer. J Back Musculoskelet Rehabil 23(1):11–19. https://doi.org/10.3233/BMR-2010-0242
8. Ciesla S, Polom K (2010) The effect of immediate breast reconstruction with Becker-25 prosthesis on the preservation of proper body posture in patients after mastectomy. EJSO 36 (7):625–631. https://doi.org/10.1016/j.ejso.2010.05.005
9. Atanes mendes Peres AC, Dias de Oliveira Latorre MD, Yugo Maesaka J et al (2017) Body posture after mastectomy: comparison between immediate breast reconstruction versus mastectomy alone. Physiother Res Int 22(1). https://doi.org/10.1002/pri.1642
10. Glowacka I, Nowikiewicz T, Siedlecki Z et al (2016) The assessment of the magnitude of frontal plane postural changes in breast cancer patients after breast-conserving therapy or mastectomy—follow-up results 1 year after the surgical procedure. Pathol Oncol Res 22 (1):203–208. https://doi.org/10.1007/s12253-015-9995-7
11. Montezuma T, Guirro ECO, Vaz MMOLL et al (2004) Changes in postural control in mastectomized women. J Cancer Ther 5(6):493–499. https://doi.org/10.4236/jct.2014.56056
12. Rachwal M, Drzał-Grabiec J, Walicka-Cupryś K et al (2013) Control of static balance among women after mastectomy. Vision impact on the quality of the equilibrium response. Postęp Rehabil 1(3):13–20
13. Gomide LB, Matheus JPC, Reis FJC (2007) Morbidity after breast cancer treatment and physiotherapeutic performance. Int J Clin Pract 61(6):972–982
14. Meneses KD, McNees MP (2007) Upper extremity lymphedema after treatment for breast cancer: a review of the literature. Ostomy Wound Manag 53(5):16–29
15. Gallagher P, Buckmaster A, O'Carroll S et al (2009) Experiences in the provision, fitting and supply of external breast prostheses: findings from a national survey. Eur J Cancer Care 18 (6):556–568. https://doi.org/10.1111/j.1365-2354.2007.00898.x

16. Glaus SW, Carlson GW (2009) Long-term role of external breast prostheses after total mastectomy. Breast J 15(4):385–393. https://doi.org/10.1111/j.1524-4741.2009.00742.x
17. Hart S, Meyerowitz BE, Apolone G et al (1997) Quality of life among mastectomy patients using external breast prostheses. Tumori 83(2):581–586
18. Hojan K, Manikowska F, Molinska-Glura M et al (2014) The impact of an external breast prosthesis on the gait parameters of women after mastectomy. Cancer Nurs 37(2):30–36. https://doi.org/10.1097/NCC.0b013e3182919576
19. Blomqvist L, Stark B, Engler N et al (2004) Evaluation of arm and shoulder mobility and strength after modified radical mastectomy and radiotherapy. Acta Oncol 43(3):280–283
20. Jemal A, Siegel R, Xu J et al (2010) Cancer statistics, 2010. CA Cancer J Clin 60(5):277–300. https://doi.org/10.3322/caac.20073
21. Gonzalez-Angulo AM, Morales-Vasquez F, Hortobagyi GN (2007) Overview of resistance to systemic therapy in patients with breast cancer. Adv Exp Med Biol 608:1–22
22. Faggiano F, Partanen T, Kogevinas M et al (1997) Socioeconomic differences in cancer incidence and mortality. IARC Sci Publ 138:65–176
23. Garvican L, Littlejohns P (1998) Comparison of prognostic and socio-economic factors in screen-detected and symptomatic cases of breast cancer. Public Health 112(1):15–20
24. Downing A, Prakash K, Gilthorpe MS et al (2007) Socioeconomic background in relation to stage at diagnosis, treatment and survival in women with breast cancer. Br J Cancer 96 (5):836–840
25. Schrijvers CT, Mackenbach JP, Lutz JM et al (1995) Deprivation and survival from breast cancer. Br J Cancer 72(3):738–743
26. Instituto Nacional do Câncer (Brasil) (2015) Estimativa 2016. Incidência do Câncer no Brasil. Rio de Janeiro: INCA
27. Advanced breast cancer: diagnosis and treatment (2009) National Collaborating Centre for Cancer, Velindre NHS Trust, Cardiff, Wales, 1–122
28. A woman's guide to breast cancer diagnosis and treatment (2000) Developed by the California Department of Health Services Breast Cancer Early Detection Program, 1–32
29. Smith RA, Saslow D, Sawyer KA et al (2003) American cancer society guidelines for breast cancer screening: update. CA Cancer J Clin 53(3):141–169
30. Plesca M, Bordea C, Houcheimi B et al (2016) Evolution of radical mastectomy for breast câncer. J Med Life 9(2):183–186
31. Hu E, Alderman AK (2007) Breast reconstruction. Surg Clin N Am 87:453–467
32. Petit J, Rietjens M, Garusi C (2001) Breast reconstructive techniques in cancer patients: which ones, when to apply, which immediate and long term risks? Crit Rev Oncol Hematol 38 (3):231–239
33. Crosible J, Kilbreath SL, Dylke E et al (2010) Effect of mastectomy on shoulder and spinal kinematics during bilateral upper-limb movement. Phys Ther 90(5):679–692. https://doi.org/10.2522/ptj.20090104
34. Devoogdt N, Van Kampen M, Christiaens MR et al (2011) Short- and long-term recovery of upper limb function after axillary lymph node dissection. Eur J Cancer Care (Engl). 20(1):77–86. https://doi.org/10.1111/j.1365-2354.2009.01141.x
35. Raaff CAL, Derks EA, Torensma B et al (2016) Breast reconstruction after mastectomy: does it decrease depression at the long-term? Gland Surg 5(4):377–384. https://doi.org/10.21037/gs.2016.05.02
36. Haddad CAS, Saad M, Perez MDCJ et al (2013) Assessment of posture and joint movements of the upper limbs of patients after mastectomy and lymphadenectomy. Einstein 11(4):426–434
37. Hojana K, Manikowskab F, Chenb BP et al (2016) The influence of an external breast prosthesis on the posture of women after mastectomy. J Back Musculoskelet Rehab 29 (2):337–342
38. Winter DA (2005) Biomechanics and motor control of human movement, 3rd edn. Wiley, Hoboken

The Role of Physiotherapy in Female Breast Cancer

Soraia Cristina Tonon da Luz and Gesilani Júlia da Silva Honório

Abstract Breast cancer is a prevalent disease in women and complications following breast cancer treatment are frequent, compromising health condition of this population. This chapter describes physiotherapy applied to motor and functional rehabilitation in women with breast cancer. The implementation of therapeutic activities is important for prevention and health promotion of these women, as for treatment, when changes are installed. Physiotherapy aims to assist the patient recovering the skills to perform daily life activities, wellness and quality-of-life during the whole treatment. It will present the main details for adequate evaluation, care and couseling, and proper physiotherapy to prevent pre- and post-operative complications.

Keywords Physiotherapy · Breast cancer · Symptom-management

S. C. T. da Luz (✉) · G. J. da Silva Honório
Department of Physiotherapy, Santa Catarina State University, Florianópolis, Brazil
e-mail: soraia.luz@udesc.br

G. J. da Silva Honório
e-mail: gesilani.honorio@udesc.br

S. Brandão et al. (eds.), *Women's Health and Biomechanics*,
Lecture Notes in Computational Vision and Biomechanics 29,
https://doi.org/10.1007/978-3-319-71574-2_6

Highlights

- Breast cancer management must include integral approach to health and involve multidisciplinary discussions about complications and comorbidities
- Efficient physiotherapy demands knowledge about associated morbidity in breast cancer at pre and post-operative phases
- Physiotherapy must be based on correct evaluation for the adequate individual demands for health conditions and couseling to patient of best practices at home

1 Introduction

Female breast cancer has high incidence, morbidity and mortality [1, 2]. Every year, 1.4 to 1.7 million new cases of breast cancer are reported, which represent 25% of all cancer in women [1, 2].

Multidisciplinary studies in this field must be continuously developed to protect women health. Because they need to consider woman's living process from the moment that cancer is found, multidisciplinary teams should be working in the diagnostic and in all treatment phases.

The physical therapist has an important role by preparing woman for medical intervention and reducing any physical constraints after cancer surgery; assisting the process of breast reconstruction; and staying by her side at any cancer recurrence and other care. Preventive actions should occur before and after cancer surgery and must consider other common treatments, such as chemotherapy, radiotherapy, hormone therapy and breast reconstruction. According to Gomide et al. [3], it is fundamental to begin rehabilitation as soon as possible, while patients still do not have constrains in range of motion, pain, lymphedema or wound adesion, because late treatment for those conditions is less effective.

The main issues after surgery are pain, reduced shoulder range of motion (ROM), lymphedema or lymphatic cording, wound complications, reduced muscle force, postural changes and lack of sensibility [1]. Thus, physiotherapy is important for motor rehabilitation to enable performing daily life activities, and to improve wellness and life quality.

In this chapter, we will discuss the physiotherapy approach to breast cancer in the pre- and post-surgical phases, with emphasis in evaluation, care and orientation about the main consequences of surgery, as well as the main physical therapy procedures applied.

2 Physiotherapy in Women with Breast Cancer

Physiotherapy must be guided by the physical condition and type of treatment. According to Hayes et al. [4], 10 to 60% of women complain about physical constraints after 6 months to 3 years of surgery. To be able to perform effective treatment and to improve women's health and life quality, it is important to know the physical and functional changes occuriing in this process [3, 4].

The following topics will discuss the process of surgical recovery, as well as the assistance aimed at the prevention and treatment of changes to this condition.

2.1 Role of Physiotherapy in the Pre-operative Phase

Pre-operative actions are dedicated to evaluating the overall clinical condition, counseling to prepare the patient for surgery, and to plan for the following process after surgery. Therefore, it is important to identify any previous morbidity in order to allow decreasing the risk of surgical complications [5–7].

Petito and Gutiérrez [5] support the relevance of starting physiotherapy in the pre-operative phase for evaluation, explanation of the treatment goals, and counseling, and should continue it as soon as possible after surgery. Besides that, Góis et al. [8] showed that physiotherapy applied into the preoperative phase improves patient's condition, improving the ROM after surgery with more shoulder abduction in 15 days and allows more functionality and independence.

During the evaluation, the physiotherapist must: evaluate trunk and upper limbs to search for skin changes and deformities; apply muscle function tests; check the shoulder for joint ROM; verify changes in body volume and breathing patterns. It is also important to screen for smoking habits, previous surgeries, chronic diseases such as hypertension or diabetes, and if she does daily-life activities independently [9].

It is necessary to counsel about how movements should be performed after surgery: up to stitches removal, maximal amplitude should initially not exceed 90°, although ROM must be gradually increased [9]. Another important recommendation concernes to breathing exercises using diaphragmatic breathing, and also how patients should lay in bed, encouraging for supine position with upper limbs at 45° under support, and when lateral recumbent, it must be at the surgery contralateral side. After breast reconstruction surgery, it is necessary to consider which sort of surgery was applied. For example, if the reconstruction included using skin above the *rectus abdomins* muscles, the patient should lay down on supine position with head elevation and small hip and knee flexion [9].

Counseling about skin hydratation and how it can be hurt (pushing and clipping tissues around the nails, axillary hair removal, limb mechanical compressions, injections, burns and others) are important to women under lymphadenectomy. It is emphasized that prior surgical plan can become different, so the patient and the professional must be sensitive to these changes, understanding that these guidelines may not be necessary for those women who do not realize axillary dissection.

2.2 Physiotherapy in the Post-operative Phase

In this phase, women experience constraints in the range of motion in the ipsilateral joints, more frequently in radial types of surgeries (such as those that are involved with the removal of the gland and, in some cases, with the removal of muscle tissue), sugeries for cancers in advanced stages, and whenever axillary lymph node dissection is necessary [10–13]. Shoulder abduction, flexion and external rotation are the most affected movements and the prescribed exercises aim to overcome such limitations [4, 11, 13].

Imobility due to pain or post-operative fear, reduced shoulder muscles force, and reduced muscle function due to neural changes such as neuropraxia and neurotmese are frequently observed [14]. Gouveia et al. [15] found decreased muscle force in the ipsilateral side after surgery, especially for the trapezius and the supraespinous muscles. Shamley et al. [16] used magnetic resonance imaging (MRI) and electromyography in such patients, and found reduced volumes for the *major* and *minor pectoralis* muscles and less active trapezius. These authors also demonstrated that mastectomy with axillary and trunk radiotherapy was associated with less trunk muscles force [16]. Thses findings may be explained by the fact that, in cancer survivors, protein sinthesis seems to be reduced due to physical inactivity, along with less aminoacids availability and increased protein degradation. There is evidence of decreased muscle cross-section area and extensibility, and also less proteins for metabolise, mainly oxidative enzimes, which may be the cause for reduced muscle oxidative potential [17].

The problems in mobility may be related to late physiotherapy. A study revealed that when the follow up occurred, on average 2.8 years after surgery, the women already had important shoulder restrictions and scapulohumeral moviments changes [18]. In this context, appropriate assessment of ROM is required, which can be done associated to the performamnce of the daily-life activities [14]. Moreover, questionnaires and functional scales can be applied to evaluate shoulder mobility and functionality, such as the Constant Shoulder Score [10], the Shoulder Pain and Disability Índex [16], the Upper Limb Disability Questionnaire [6] and the Disabilities of the Arm, Shoulder, and Hand (DASH) survey [19].

Inclinometers [20] and goniometers [6, 11–13, 15, 16, 18, 21] also should be used to evaluate shoulder, elbow and wrist positions and ROM in order to compare joints or sides. Besides that, muscle force can be evaluated by dinamometers [10, 20] and manual muscle testing grades [6, 15, 19], which can be associated to changes in joint ROM.

Early intervention seems to be benefitial to recovery the joint´s ROM, through proper exercises that enhance functionality [3, 4, 6, 9]. Physiotherapy is benefitial to recover the biomechanical action and performance of the shoulder increasing the ROM to over 90° after the stitches and drains are removed [9]. These exercises are prescribed according to the women´s personal needs at the post-operative phase [5, 6, 8, 13]. Importantly, if women still have the stiches and drains, ROM movements at 90 should be restricted [9].

Moderate-intensity aerobic exercises are recommended [17, 22] along with stretching exercises for the upper limb, neck and upper trunk [3, 5, 7, 9], emphasizing the scapular movements to avoid adhesions. In case of winged scapula, exercises with the *serratus* anterior are also suggested [3, 9]. Mobility exercises and strength training can help recovering the ROM, muscle function and stop or reduce impairments in the daily-life activities [6, 19]. Stan et al. [21] showed that Pilates exercises has increased shoulder abduction and rotation in women with breast cancer in post-operative phase. Changes in lymph volume at the axilla and arms at an imediate post-operative phase are frequent when drainage was removed, as in the presence of seroma and infections, under highly invasive surgeries and when the invaded lymph nodes are removed [11–13, 23]. Axillary lymph node dissection defines counseling and physiotherapy intervention. Lymphedema at the ipsilateral limb is fully related to axillary lymphadenectomy. Axillary radiotherapy also deads to lymphodema because radiation might block axillar lymph circulation [12, 13]. The fear to move the arms, which is declared by many women, may aggravate the lymphedema due to the reduced lymph circulation [24]; therefore, physiotherapists should give the correct counsiling about this issue.

Physiotherapy applied to lymphedema conditions is focused on both prevention and treatment to reduce the volume of the upper limb [13, 23, 25, 26]. The evaluation starts with anthropometric measures, using the olecranon or the wrist as references for circunference measurements [20, 23], and the difference between sides as a result of lymphedema should be above two centimeters [9]. Limb volume also can be evaluated by measuring liquid displacement [11, 25] or by limb volume estimation [12, 19].

To prevent lymphodema, it is important to start the exercises of the upper limbs immediately after the surgery as well to keep the ipsilateral upper limb in higher position than that of the torso [19, 20, 23]. Self-massage is also recommended for radical axillary lynph node dissection, by stimulating contralateral lymph nodes and massaging from the ipsilateral axilla to the opposite side [9, 27]. Complete descongestive therapy applies exercises, manual lymphatic drainage, compression gairnments and skin care to succesfully recover lymphedema [26–28], which can also prevent lymphodemaIt important to use the most appropriated exercises for lymphedema treatment [13, 19]. Rezende et al. [23] and Sagen et al. [25] showed that increased ROM does not increase lymphatic complications, and also that imobility does not protect against lymphedema. IIt is fundamental to treat physical and social consequences of lymphedemas [29] because its morbidity is related to pain, motion reduction, skin changes, and neurologic disfunction. Social and behavioral conditions can be negatively stressed because women's frailty and anguish, affecting her perception of the quality of life [29].

Axillary web syndrome or lymphatic cording sometimes develops in post-operative phase, when ropelike structures are felt under the skin when shoulder moves in abduction or external rotations [19]. Such lumps might go down to the elbow, wrist or fingers, causing pain and disconfort and constraing the ROM. Early physiotherapy is helpfull to prevent cording by using stretching exercises and moviment of the affected joints, as well as manual therapy upon the ropelike structure and lymphatic drainage [6, 10]. Manual lymphatic drainage improves blood flow and increases lymphatic circulation to avoid axillar blockage that can generate the cording. Drainage along with active and passive exercise, stretching upon the ropelike structure and shoulder mobilization also reduce or eliminate it [19, 26, 27]. Breast cancer surgery may affect the sensibility close to surgery location. If one or more nerves were affected, their dermatome will also evidence changes in sensibility [7]. This lack of sensibility increases the risk for injuries, infections and inflamatory processes that will worsen the mobility and quality-of-life. Changes in the sensibility of the trunk affect postural orientation and might affect reactive postural actions and postural control [30].

During axillary lymphadenectomy, the injury of intercostobrachial sensitive nerve leads to loss of sensibility in the upper arm and axilla [31, 32]. Madsena et al. [11] found 70% of women presenting lost or reduced sensibility after primary axillar dissection surgery, and 22% presented the same condition after sentinel lymph node biopsy. Pain and parestesis were the most common symptons observed.

Monofilament test for sensibility is recommended at the axillar region and upper limb dermatomes, mainly close to intercostobraquial nerve and to the affected breast, which should be divided into 4 areas [31, 32]. This evaluation should compare sides and be applied along and after the treatment.

Before treatment, it is necessary to evaluate the level of sensibility and what is the kind of change in sensibility: hypoesthesia, hyperesthesia or lack of sensibility [30–32]. Resensibility techniques using different textures and temperatures can be used. In absence of graphesthesia, it is possible to write down with gross pencil at the affected areas. Under stereognosis, one should draw geometric forms and ask the patient to figure out what is the draw with her eyes closed.

Exercises for upper limb are also important to recover sensibility, as as well resensibilization and manual lymphatic drainage to stimulate sensory receptors and improve the skin sensibility threshold [3, 7, 9].

Phantom feelings and pain may ocurr in the imediate post-operative phase when the patient declares that she is still feeling or having pain in the amputated limb. In the phantom breast syndrome, phantom feeling and/or pain may occur in the whole or just in part of the phantom breast. The phantom pain is variable in intensity, duration and type, as well as in its prevalence up to 30% of mastectomized women [33, 34]. However, we should notice that it is usually under reported because such symptoms are not screened during evaluation.

Due to nerve dissection and organ removal, neural reorganization takes place. As dynamics of pain has sensory, physical and environmental aspects, a biopsicossocial approach is adviced along the treatment, starting by an evaluation woth the McGill questionnaire [35].

The Mirror Technique should be used with caution and exercise should be gradually performed, as more parts of the body are included, from still trunk to upper limb and trunk exercises. The use of external prosthesis might assist the motor-sensory reorganization. Therapies that include body expression are useful, although medication may be useful for pain control may be applied together [36].

Postmastectomy pain syndrome (PMPS) is a chronic neuropathic pain disorder that can occur following breast cancer procedures at the homolateral upper limb, chest and axilla, which lasts for more than 3 months. PMPS induceb by intercostobrachial nerve injury is the most common situation, but the exact symptoms may vary acording to the injured nerve, in this case more paresthesis than pain is observed [37, 38].

MacDonald et al. [37] freported that 52% of women (5–7 years after surgery) had PMPS, while Springer et al. [6] found that 60% of post-operative women had PMPS that decreased spontaneously in few cases, which emphasisis the need for intervention against PMPS.

There are different types of neurophatic pain following breast cancer surgery [38]: (i) breast phantom pain; (ii) intercostobrachial neuralgia: pain and sensitive changes related to intercostobrachial nerve distribution; (iii) pain secondary to the presence of a neuroma; and (iv) pain due to damage to other nerves.

Stretching exercises, muscle relaxation and active and assisted-active upper limb exercises help to reduce pain. Cervical relaxation and cervical and scapular mobilization are also too [9, 25, 38]. Massotherapy at the wound site can improve skin condition and reduce pain, as well lymphatic self-massage [9].

Electrotherapy (transcutaneous nerve stimulation and interferential current therapy) should be used only when there is no loss of local sensibility or any other contraindication [9, 38].

Wound and skin changes might happen after breast cancer treatment. Main complications for those conditions are dehiscence, infection, skin adherence and keloid. These complications are usually observed at the incision site, donor site for skin graft to breast reconstruction, drain site and medial site of axillar radiation therapy [18].

Infection and dehiscense are not affected by the kinesiotherapy-controlled angulation [23]. Therefore, exercises should only exceed the 90º after the removal of the stitches and drain, and only then the shoulder ROM should be progressively increased. To avoid dehiscence, the joint ROM during exercises cannot provoke wound tension. Upper limb stretching exercises can reduce skin adherence and must always consider patient's post-operative conditions [9, 23].

After removing the stitches and at the end of proper cicatrization, some manouvers such as deep transverse massage can be applied over the surgical wound to reduce the skin adesion [9]. Those procedures are also available for wound problems like hypertrophy or keloide.

Mastectomy can affect breathing, mainly due to surgery location, imobility in bed and fear to move and breath correctly. Moreover, intubation during surgery can cause lung secretion accumulation and the patient might be afraid to cough and expel it [39]. Postures to improve breathing function when lying in bed should be adopted for the early post-operative phase, e.g., the semi-Fowler position to improve dynamical breathing and diafragm function [9, 27]. If necessary, the huffing technique and assisted coughing can be applied to clear the lungs as support to breathing exercises for lung expansion and diafragma education, with or without incentive spirometry [9]. In this context, it is always important to verify the smoking habits, or any respiratory disease such as asma, and also if the patient was previously subjected to any of these respiratory-related treatments.

Antalgic postures are the main postural changes after breast cancer surgery: kyphosis, forward head or flexing the elbow close to the trunk [40]. For postural rehabilitation, it is important to improve fitness—by including exercises for shoulder mobility and strength–, for postural correction and to develop postural stability, breathing exercises, and exercises to prevent lymphedema [3, 5, 7, 14, 40]. Exercises to increase active joint ROM at the operated side must come along with body perception, and it can be done by using postural reeducation and Pilates exercises [21]. Postural recommendations such as body posture alignment during standing and walking and loosing the arms are important. External breast prosthesis reduces trunk deviations. Rostkowska et al. [40] showed that despite mastectomy increases cyphosis, using external breast prosthesis and physiotherapy sessions decrease such postural problem.

The patient should avoid overloading the shoulder and arm at the operated side mainly during the first 30 days after surgery. Trunk and shoulder exercises should be carefully taken when establishing the traning volume, i.e., load and number of repetitions. It must begin after post-operarive complications were overcame, such as lymphodema, seroma, cicatrization problems, lymphatic cording and muscle fatigue [3, 9, 40].

A healthy sexual life is important for cancer survivors. Women may experience some sexual problems after breast cancer treatment, such as reduced desire and sexual arousal, and lack of vaginal lubrification, leading to dyspaurenia and difficulty to reach the orgasm. This can be triggered by hormonal treatments after radiation therapy, which may cause vulvovaginal atrophy, dyspaurenia and low libido [41, 42]. The physiotherapist should also focus on evaluating the pelvic floor muscles and any sexual health condition, such as vaginal lubrification, scars, stenosis or genital flaccity. The Oxford Scale [43] and the PERFECT scheme [44] are two easy-to-apply assessment tools, as well as the Female Sexual Function Index (FSFI) questionnaire. This survey assesses sexual function in women, by dividing it into six dimensions: sexual drive, excitation, lubrification, orgasm, satisfaction and pain, and can be applied to evaluate the effects of the physiotherapy [43, 45].

These sexual conditions can be treated using manual therapy, perineal muscle massage to induce relaxation, trigger point release, proprioception exercises by using finger touch to induce fast and slow muscle contractions. Vaginal weights also can be used for proprioception and muscle training. Pelvic floor muscle training can be performed together with upper limb exercises. For vaginal stenosis, vaginal dilator can be used [45, 46].

Cognitive disfunction is a side effect observed in adjuvant breast cancer treatment (chemotherapy) or in cancer surgery. While the chemotherapeutic agents can have a pysiologics effect on the Central Nervous System, stress and anxienty may be triggered by the disagnosis and treatment, which changes the way of living and diminishes the quality-of-life [47].

A sistematic review published by Morean, O′Dwyer and Cherney [48] showed that the main resources to deal with cognitve changes in cancer are medication, resistance training (whenever the patient is cooperative), and restorative and cognitive treatments. Within restorative treatment, meditation and relaxation exercises are included. Activities such as balance training, body image exercises, attention and memory exercises, as well relaxation may be used as cognitive treatments [48].

Palliative care shoud be a humanist and integrate approach taht includes psychological, social and spiritual support to breast cancer patients with life-threating risk or without cure, aiming to minimize cancer symptoms and improve the quality of life [49]. Several physiotherapy approaches are available to reduce pain, generate relaxation to face depression and stress, and to deal with orthopedic conditions, using stretching exercises, relaxation and resistance exercises to overcome fadigue [50, 51]. Improving respiratory function by means of breathing exercises is essencial. Problems related to lymphodema and lymphatic cording can be treated with manual lymphatic drainage, manual streyching of lymphatic structures and exercises [26, 51].

3 Conclusions

Physiotherapy is essential in the rehabilitation of breast cancer in women. Anticipation of clinical complications is mandatory to everyone involved in breast cancer prevention, providing the most suited couseling to patients and family to avoid unecessary risks. Choosing the physical therapy protocols must respect and consider global evaluation and any process-related complication. Using physical therapy without clearly set the goals brings forth unsafety, and therefore reduces patient's trust and treatment success. It is necessary to go towards improving the quality-of-life through relief of symptons and improving functional independence.

References

1. Holsbach N, Fogliatto FS, Anzanello MJ (2014) A data mining method for breast cancer identification based on a selection of variables. Ciênc Saúde Colet 19(4):1295–1304
2. Woolston C (2015) Breast cancer. Nature 527(7578):101
3. Gomide LB, Matheus JPC, Reis FJC (2007) Morbidity after breast cancer treatment and physiotherapeutic performance. Int J Clin Pract 61(6):972–982. https://doi.org/10.1111/j.1742-1241.2006.01152.x
4. Hayes SC, Johansson K, Stout NL et al (2012) Upper-body morbidity after breast cancer. Cancer 15:2237–2249. https://doi.org/10.1002/cncr.27467
5. Petito EL, Gutiérrez MGR (2008) Development and validation of an exercise program for women submitted to breast cancer surgery. Rev Bras Cancerol, 2008. 54(3): 275–287
6. Springer BA, Levy E, McGarvey C et al (2010) Pre-operative assessment enables early diagnosis and recovery of shoulder function in patients with breast cancer. 120(1):135–147. https://doi.org/10.1007/s10549-009-0710-9

7. McNeely ML, Binkley JM, Pusic AL et al (2012) A prospective model of care for breast cancer rehabilitation: postoperative and postreconstructive issues. Cancer 15:2226–2236. https://doi.org/10.1002/cncr.27468
8. Góis MC, Furtado PR, Ribeiro SO et al (2012) Range of motion and measure of functional independence in patients with mastectomized axillary dissection. Rev Ciênc Méd 21(6):111–118
9. Bergmann A, Ribeiro MJP, Pedrosa E et al (2006) Physical Therapy in Breast Cancer: clinical protocol at the Cancer Hospital III/ INCA. Rev Bras Cancerol 52(1):97–109
10. Lauridsen MC, Christiansen P, Hessov I (2005) The effect of physiotherapy on shoulder function in patients surgically treated for breast cancer: a randomized study. Acta Oncol 44 (5):423–424
11. Madsena AH, Haugaard K, Soerensen J et al (2008) Arm morbidity following sentinel lymph node biopsy or axillary lymph node dissection: a study from the Danish Breast Cancer Cooperative Group. Breast 17(2):138–147. https://doi.org/10.1016/j.breast.2007.08.006
12. Nesvold I, Dahl AA, Løkkevik E et al (2008) Arm and shoulder morbidity in breast cancer patients after breast-conserving therapy versus mastectomy. Acta Oncol 47(5):835–842. https://doi.org/10.1080/02841860801961257
13. Scaffiddi M, ulpiani MC, Vetrano M et al (2012) Early rehabilitation reduces the onset of complications in the upper limb following breast cancer surgery. Eur J Phys Rehab Med 48 (4):601–611
14. Nascimento SL, Oliveira RR, Oliveira MMF et al (2012) Complications and physical therapeutic treatment after breast cancer surgery: a retrospective study. Fisioter Pesqui 19 (3):248–255. https://doi.org/10.1590/S1809-29502012000300010
15. Gouveia PF, Gonzalez EO, Grer PA et al (2008) Shoulder motion range and strength assessment in late post-operative patients having undergone modified radical mastectomy. Fisioter Pesqui 15(2):172–176. https://doi.org/10.1590/S1809-29502008000200010
16. Shamley DR, Srinanaganathan R, Weatherall R et al (2007) Changes in shoulder muscle size and activity following treatment for breast cancer. Breast Cancer Res Treat 106(1):19–27. https://doi.org/10.1007/s10549-006-9466-7
17. Schneider CM, Hsieh CC, Sprod LK et al (2007) Cancer treatment-induced alterations in muscular fitness and quality of life: the role of exercise training. Ann Oncol 18:1957–1962. https://doi.org/10.1093/annonc/mdm364
18. Fourie WJ (2008) Considering wider myofascial involvement as a possible contributor to upper extremity dysfunction following treatment for primary breast cancer. J Bodyw Mov Ther 12(4):349–355. https://doi.org/10.1016/j.jbmt.2008.04.043
19. Cho Y, Do J, Jung S et al (2016) Effects of a physical therapy program combined with manual lymphatic drainage on shoulder function, quality of life, lymphedema incidence, and pain in breast cancer patients with axillary web syndrome following axillary dissection. Support Care Cancer 24(5):2047–2057. https://doi.org/10.1007/s00520-015-3005-1
20. Merchant CR et al (2008) Decreased muscle strength following management of breast cancer. Disabil Rehabil 30(15):1098–1105
21. Stan DL, Rausch SM, Sundt K et al (2012) Pilates for breast cancer survivors: impact on physical parameters and quality of life after mastectomy. Clin J Oncol Nurs 16(16):131–141. https://doi.org/10.1188/12.CJON.131-141
22. Travier N, Velthuis MJ, Steins Bisschop CN et al (2015) Effects of 18-week exercise programme started early during breast cancer treatment: a randomised controlled trial. BMC Med 121(13):1–12. https://doi.org/10.1186/s12916-015-0362-z
23. Rezende LF, Beletti PO, Franco RL et al (2006) Random clinical comparative trial between free and directed exercise in post-operative complications of breast cancer. Rev Assoc Med Bras 52(1):37–42. https://doi.org/10.1590/S0104-42302006000100020
24. Karadibak D, Yildirim Y, Kara B et al (2009) Effect of complex decongestive therapy on upper extremity lymphedema. Fizyoterapi Rehabilitasyon 20(1):3–8
25. Sagen S, Kåresen R, Risberg MA (2009) Physical activity for the affected limb and arm lymphedema after breast cancer surgery. A prospective, randomized controlled trial with

two-year follow-up. Acta. Oncology 48(8):1102–1110. https://doi.org/10.3109/02841860903061683
26. Lasinski BB (2013) Complete decongestive therapy for treatment of lymphedema. Semin Oncol Nurs 29(1):20–27
27. Buragadda S, Alhusaini AA, Melam GR et al (2015) Effect of complete decongestive therapy and a home program for patients with post mastectomy lymphedema. J Phys Ther Sci 27:2743–2748. https://doi.org/10.1589/jpts.27.2743
28. Castro-Sánchez AM, Moreno-Lorenzo C, Matarán-Peñarrocha GA et al (2011) Preventing lymphoedema after breast cancer surgery by elastic restraint orthotic and manual lymphatic drainage: a randomized clinical trial. Med Clin (Barc) 137(5):204–207. https://doi.org/10.1016/j.medcli.2010.09.020
29. Laird BJA, Fallon MT (2009) Palliative care in the elderly breast cancer patient. J Clin Oncol 21:131–139. https://doi.org/10.1016/j.clon.2008.11.009
30. Beahm EK, Walton RL (2006) Sensibility following innervated free tram flap breast reconstruction. Plast Reconstr Surg 117:2126–2128. https://doi.org/10.1016/S1043-321X(07)80284-0
31. Gahm J, Hansson P, Brandberg Y et al (2013) Breast sensibility after bilateral riskreducing mastectomy and immediate breast reconstruction: a prospective study. J Plast Reconstr Aesthet 66:1521–1527. https://doi.org/10.1016/j.bjps.2013.06.054
32. Bocatto AM, Haddad SCA, Rizzi SCL et al (2013) Assessment of tactile sensitivity and function of upper member in the post surgery of mastectomy compared with quadrantectomy. Rev Bras Mastologia 23(4):117–123
33. Poma S, Varenna R, Bordin G et al (1996) The phantom breast syndrome. Rev Clin Esp 196(5):299–301
34. Dijkstra PU, Rietman JS, Geertzen JH (2007) Phantom breast sensations and phantom breast pain: a 2-year prospective study and a methodological analysis of literature. Eur J Pain 11(1):99–108. https://doi.org/10.1016/j.ejpain.2006.01.002
35. Melzack R (1975) The McGill pain questionnaire: major properties and scorin methods. Pain 1(3):277–299
36. Brandberg Y, Sandelin K, Erikson S et al (2008) Psychological reactions, quality of life, and body image after bilateral prophylactic mastectomy in women at high risk for breast cancer: a prospective 1-year follow-up study. J Clin Oncol 6(24):3943–3949. https://doi.org/10.1200/JCO.2007.13.9568
37. MacDonald L, Bruce J, Scott NW et al (2005) Long-term follow-up of breast cancer survivors with post-mastectomy pain syndrome. Br J Cancer 92:225–230. https://doi.org/10.1038/sj.bjc.6602304
38. Couceiro TC, Menezes TC, Valenca MM (2009) Post-mastectomy pain syndrome: the magnitude of the problem. Rev Bras Anestesiol 59(3):358–365
39. Aaron J, Sasidharan B, Ebenezer S et al (2015) Impact of breathing on post-mastectomy radiotherapy: a dosimetric comparison between intensity-modulated radiotherapy and 3D tangential radiotherapy. J Radiother Pracct 14(2):126–134. https://doi.org/10.1017/S1460396915000096
40. Rostkowska E, Bąk M, Samborski W (2006) Body posture in women after mastectomy and its changes as a result of rehabilitation. Adv Med Sci 51:287–297
41. Kowalczyk R, Lew-Starowicz Z, Krzemieniecki K et al (2014) EPA-0916–Sexual function in women after mastectomy: a survey of polish patients. Eur Psychiatry 29:(S1). https://doi.org/10.1016/S0924-9338(14)78241-0
42. Aerts L, Christiaens MR, Enzlin P et al (2014) Sexual functioning in women after mastectomy versus breast conserving therapy for early-stage breast cancer: a prospective controlled study. Breast 23:629–636. https://doi.org/10.1016/j.breast.2014.06.012
43. Kanter G, Rogers RG, Pauls RN et al (2015) A strong pelvic floor is associated with higher rates of sexual activity in women with pelvic floor disorders. Int Urogynecol J 26:991–996
44. Laycock J, Jerwood D (2001) Pelvic floor muscle assessment: the PERFECT scheme. Physiotherapy 87(12):631–642. https://doi.org/10.1016/S0031-9406(05)61108-X

45. Tahereh E, Sohrabi M, Haghollahi F et al (2014) Comparation effect of physiotherapy with surgery on sexual function in patients with pelvic floor disorder: a randomized clinical trial. Iran J Reprod Med 12(1):7–14
46. Cacchioni T, Wolkowitz C (2011) Treating women's sexual difficulties: the body work of sexual therapy. Sociol Health Ill 33(2):266–279. https://doi.org/10.1111/j.1467-9566.2010.01288.x
47. Munir F, Wiskemann J, Armbrust P et al (2011) Cognitive intervation for breast cancer patients undergoing adjuvant chemotherapy: a needs analysis. Cancer Nus. 34:385–392. https://doi.org/10.1002/ijc.29383
48. Morean DF, O'Dwyer L, Cherney LR (2015) Therapies for cognitive déficits associated with chemotherpy for breast breast câncer: a systematic review of objective outcomes. Arch Phys Med Rehabil 96:1880–1897. https://doi.org/10.1016/j.apmr.2015.05.012
49. Pidlyskyj K, Roddam H, Rawlinson G et al (2014) Exploring aspects of physiotherapy care valued by breast cancer patients. Physiotherapy 100(2):156–161. https://doi.org/10.1016/j.physio.2014.03.006.100:15661
50. López-Sendín N, Alburquerque-Sendín F, Cleland JA et al (2012) Effects of physical therapy on pain and mood patients with terminal cancer: a pilot randomized clinical trial. J Altern Complement Med 18(5):480–486. https://doi.org/10.1089/acm.2011.0277
51. Beaton R, Pagdin-Friesen W, Robertson C et al (2009) Effects of exercise intervention on persons with metastatic cancer: a systematic review. Physiother Can 61(3):141–153. https://doi.org/10.1111/j.1467-9566.2010.01288.x

Breast Implants: Far Beyond Just Aesthetic Surgery

Nilza Ramião, Fabiana Zanata, Tomas Fortoul, Fabio X. Nahas, Lydia M. Ferreira, Pedro Martins, Antonio Augusto Fernandes, Maria da Luz Barroso and Diana Costa Santos

Abstract For decades, women have undergone breast implant surgery either for health or aesthetic reasons. Some adverse effects have made their history full of events, showing the complex factors involved in the interaction of the device with the human body. Understanding the adverse outcomes is a key factor to improve the breast implants. This will help to increase the safety of the procedure. It is also important to stress some new concerns as well as the lack of evidence of involvement on specific long-term health outcomes in women with silicone gel breast implants, such as cancer; connective tissue, rheumatologic, and autoimmune diseases; neurologic diseases; reproductive issues, including lactation; offspring issues; and mental health issues. The development of the National Breast Implant Registry (NBIR), since 2012 by the American Society of Plastic Surgeons in collaboration with the FDA was a huge step to understand these issues. In this database, it is possible to collect and tracks important information on patients and devices, maximizing data extraction from clinical cohorts and directing more meaningful analyses. Researchers could finally obtain clarification of the long-term health outcomes of silicone gel breast implants.

Keywords Breast augmentation · Mammoplasty · Implant capsular contracture
Breast implants · Adverse outcomes

N. Ramião (✉) · P. Martins · A. A. Fernandes
INEGI, Faculty of Engineering of University of Porto, Porto, Portugal
e-mail: nilza.ramiao@gmail.com

F. Zanata · F. X. Nahas
Division of Plastic Surgery, Federal University of São Paulo, São Paulo, Brazil

T. Fortoul
Plastic Surgery Unit, Centro Médico Docente La Trinidad, Caracas, Venezuela

L. M. Ferreira
Chairwoman of the Division of Plastic Surgery, Federal University of São Paulo, São Paulo, Brazil

M. da Luz Barroso · D. C. Santos
Department of Plastic Surgery, Centro Hospitalar de Gaia-Espinho—EPE, Vila Nova de Gaia, Portugal

S. Brandão et al. (eds.), *Women's Health and Biomechanics*,
Lecture Notes in Computational Vision and Biomechanics 29,
https://doi.org/10.1007/978-3-319-71574-2_7

Highlights

- The major concern to the patient and the plastic surgery community is the environmental durability and useful life of a breast implant
- The chapter presents some tips to development and improvement of safer and more compliant product
- The collaboration of all plastic surgery community manufacturers and regulatory health agencies would be crucial in the development of a database about breast implants
- The analysis of the explanted implants is the key to improvement the breast implants

1 Introduction

In the last few decades, breast augmentation has evolved significantly due to the development of new materials and technical improvement. Breast augmentation is the third aesthetical surgical procedure most performed in the world, with approximately 1.348.197 surgeries per year [1].

Cronin and Gerow have developed the first silicone breast prosthesis in 1961. Its first use for breast augmentation was in 1962 [2].

The basic design of silicone breast implants is silicone rubber envelops or shells filled with silicone gel of various molecular weights. Medical grade silicone is a specific polymer called poly(dimethylsiloxane) (PDMS), and the more units in its formation, the more solid the silicone polymer becomes. Low-molecular-weight silicone polymers are oils (fluids), whereas higher-molecular-weight silicones are more viscous. The silicone gel used in breast implants contains a mixture of both low and high molecular weight polymers in various ratios. The low molecular weight silicone fluid is not chemically attached to the main gel and can move or be extracted the same way as water in a sponge [3] The implant shell can be single or double, smooth or textured, or covered with polyurethane foam. Shells vary according to the composition and characteristics of the elastomer, the type of coating, the number of layers, and the type of barrier layer used [4]. Barrier layers are specifically constructed to lessen the diffusion of low-molecular-weight silicone fluids into the surrounding tissues (gel leak or bleed). Despite these barrier layers, Yu et al. [5] measured the silicone gel fluid diffusion of various implant models at about 300 mg per year, with considerable variation depending on implant age and manufacturer.

During almost 30 years of silicone gel breast implants usage, there were several reports of perceived and suspected clinical problems leading to many lawsuits; this led the Food and Drug Administration (FDA) to launch a *moratorium* in 1992 in the United States [6], announcing that silicone gel-based breast implants would only be available, through controlled clinical studies, to which only women who needed implants for breast reconstruction, would have access [4]. The next year, FDA also decided that saline filled implants needed more studies. Succeeding these announcements, there has been an increase in number of the research concerning the behavior of silicone breast implants to provide a rationale for more realistic risk–benefit assessments by plastic surgeons and patients [7–9]. After extensive studies and analysis carried out by companies such as Mentor, Inc and McGhan (formely Inamed), the FDA declared the implants safe in 2006, but imposed some restrictions such as minimum age for implants, patient tracking and requested more information should be provided to patients [3, 10, 11]. Despite FDA reports breast implants kept being manufactured and improved to the present day. The initial breast implants designed by Cronin and Gerow and produced by Dow Corning had a thick smooth shell and a thick silicone gel roughly consisting of 50% low molecular-weight silicone fluid and 50% high molecular-weight silicone gel. The second-generation implants changed to thinner shells and softer gels (80% low-molecular-weight silicone fluids and 20% high-molecular-weight silicone gel) to provide a more natural feel, but sooner it became clear that increasing gel bleed and rupture rates occurred. The third-generation implants introduced barrier layers and increased the cohesiveness of the gels to reduce gel bleed. Fourth-generation implants introduced shell texturing to reduce capsular contracture and increased gel cohesiveness to improve the feel and performance of the implants. Finally, fifth–generation implants have a form-stable design with increased gel cohesiveness and an anatomic shape [12].

The concerns about safety and durability of silicone breast implants did not stop. The life span of silicone implants was initially presumed to be unlimited, but it was later demonstrated that the silicone elastomer has a finite life span and that silicone implants age and eventually fail. Rohrich et al. [13] reported implant failure rates of 4–71% depending on the definition of implant failure, the population, and the diagnostic method used. According to the Inamed and Mentor Silicone Breast Implant Core Studies published in 2007, the overall implant rupture rate was 3.5% at 6 years for Inamed implants and 0.9% at 3 years for Mentor implants [10, 11].

In March 2010, the French medical device regulatory agency (AFSSAPS) suspended the marketing, distribution and use of all silicone implants produced by Poly Implant Prostheses (PIP) due to serious concerns about the quality of the material applied [14]. The recent controversy surrounding the PIP breast implants caused heightened anxiety and extensive publicity regarding their safety in humans. Based on peer-reviewed published studies the probability of rupture for PIP implants is estimated to be around 25–30%, 10 years after implantation, whereas other silicone breast implants have been reported with a rupture rate of 2–15% considering the same follow-up period [15]. New studies have been developed since 2010 until now, to understand the causes of rupture [15–17].

In this context, the aims of this chapter are: (i) enumerating the most common issues/complications on the safety of breast implants (ii) identifying the local complications and adverse outcomes that may occur in patients at any time after breast implant surgery. Finally, based on our experience and through the regulatory health agencies report, we try to answer to the question: "How to improve safety of breast implants?".

2 Understanding Adverse Outcomes to Improve Breast Implants

We will summarize the main adverse outcomes of breast implants, in order to recommend the safety of breast implants, based on the review of the literature and the author's experience. The problems that may occur in patients with breast prosthesis can be local (capsular contracture, implant displacement, breast pain, breast tissue atrophy, infection, seroma) or device-related problems (rupture). In any case, the need for additional surgery, with or without removal of the device, or other medical interventions might be necessary.

The inclusion of prostheses to increase breast volume is an aesthetic based surgery and with this in mind it is important to consider that the final outcome will depend on the patient's anatomy, surrounding tissues and the placement of the implant. The most common adverse outcomes are capsular contracture, malposition or displacement, and rupture or deflation of the implant [18, 19]. Other less frequent problems include wrinkling, asymmetry, scarring, pain, seroma and infection.

Capsular contracture is a normal reaction of the body to a foreign agent. A protective capsule of fibrous tissue is formed around the intruder, resembling an articular synovia producing liquid on periprotetical space [20]. There are four grades of capsular contracture, known as Baker Grades, according to the severity of the problem (in grade I, the breast is normally soft and looks natural, and in grade IV the breast is hard, painful, and looks abnormal). According to the Institute of Medicine committee, there is evidence that saline-filled implants have a reduced rate of contracture compared to implants filled with silicone; however, these are not conclusive studies [18, 21].

Also, textured-surface implants have been associated with less intense or less frequent capsular contraction compared to those with smooth surfaces (15% *vs* 9%, respectively) [22]. This evidence can be explained by the fact that textured and microtextured implants were designed to better adhere to the breast tissue and this adherence should prevent movement or rotation of the implant and it that way reduce the capsular contracture [23]. The prosthesis location, submuscular, subglandular or subfascial, is also important to reduce the risk of capsular contracture. Studies have demonstrated that the rate of capsular contracture as low as 10% with submuscular surgery compared to 30% with implants placed in the subglandular position [18, 21]. Also, polyurethane coated gel filled implants were used in US until its withdraw from the markets in 1990, but they have been still marketed in Europe and Latin America (Politech Implants, Germany and Silimed, Brazil). There were recently reports that have associated these implants to a low capsular contracture incidence rates. Polyurethane coated implants may be efficient in reducing capsular contracture rates or revision rates; however, the available studies do not show whether there is a real difference among the implant types [24].

The implant displacement is another local impaired outcome after breast augmentation surgery. The ideal situation is to achieve a definite result at the end of the surgery, but as a living tissue it changes over time. Some implants progressively to increase poket space of the pocket, lowering (bottoming down), becoming more lateral or even flipping over itself. This situation has been described as capsular weakness, which is the expansion of the capsule due to gravitational forces acting over the implant on a weak capsule, leading to these complications [25].

Breast tissue atrophy also demands some attention. The device makes a prolonged compression on the skin, breast tissue and muscle producing a reduction on its final volume. The reduction in volume due to breast tissue atrophy tends to be more intense in cases of subglandular prosthesis as compared to submuscular [26].

Implant rupture may be secondary to many factors, such as: in vivo degradation of the shell mechanical properties with the exposure to a biological environment; the inadvertent instrument damage during surgery; needle biopsy or hematoma aspiration; the compression during a mammography; physical stresses to the breast (trauma or intense physical pressure); shell wrinkling; patch detachment, manufacturing defects (quality problems) and fatigue. Rupture can cause deflation when filer material leaks from the breast implant. Release or migration of silicone into the surrounding tissues induces a nonspecific foreign body reaction, resulting in typical macrophage invasion, giant cell formation, and eventual scarring [3, 27, 28].

Silicone lymphadenopathy is a deposition of silicone in one or more lymph nodes due to lymphatic migration of silicone and represents a normal physiologic response to the presence of foreign material [28]. Silicone seems to be phagocytosed by multinucleated giant cells that have uniformly distributed nuclei with abundant eosinophilic cytoplasm and prominent vacuoles partly containing silicone gel [17]. When it occurs, silicone lymphadenopathy typically is a late finding and can cause considerable morbidity and anxiety to patients with enlarged axillary lymph nodes [28].

In extreme cases, women can have local infection, pain, atrophy, inflammation/irritation, necrosis, calcifications, etc. These problems require medical treatment, additional surgery, and possibly replacement, a temporary or permanent removal of the implant [29].

Due to this, in case of implant rupture and/or in the presence of loco-regional inflammation or lymphadenopathy, there is general agreement among plastic surgeons that explantation is needed, followed by observation of the lymphnode chain.

Increasing concern has been raised for systemic problems in patients with breast implants. The anaplastic large cell lymphoma (ALCL) is a rare type of lymphoma that could be related to the silicone filled implants. Balk et al. [30] systematically reviewed the literature regarding specific long-term health outcomes in patients with silicone gel breast implants, including cancer; connective tissue, rheumatologic, and autoimmune diseases; neurologic diseases; reproductive issues, including lactation; offspring issues; and mental health problems (depression and suicide). The evidence remains inconclusive about any association between silicone gel implants and long-term health outcomes. Better evidence is needed, from existing large studies, which can be reanalyzed to clarify the strength of associations between silicone gel implants and health outcomes [30].

3 How to Improve Safety of Breast Implants?

For decades, women have undergone breast implant surgery either at a health or aesthetic level. "Overall, around 80% of all breast implantations are performed for cosmetic reasons and about 20% for reconstructive purposes (cancer)" [8]. For that reason, the questions about the safety of breast implants materials are constantly appearing. The major concern to the patient and the plastic surgery community is the environmental durability and useful life of a breast implant.

By approaching the main topics that require more research, we will provide a development and improvement of safer and more compliant products as well as be a significant tool in the future scientific research and product monitoring. The recommendations for further works were selected according to the basis of our experience with material behavior of PIP breast implants, and through the regulatory health agencies reports [14, 18, 29, 31].

The informed consent to all patients regarding all aspects of the inclusion of silicone breast prosthesis including possible risks, information about the specificities of their breast implant (such as, brand, shape, surface, size, among others)—is key factor to improve general population on the knowledge and expectations to reality. One of the recommendations is the need for standard procedures in informing women about breast implants, conducting studies on their safety, and approving changes in design [18].

According to the Scientific Committee on Emerging and Newly Identified Health Risks, the implementation of a registration system of breast implants on a national, European or International level would be relevant. The aim would be collecting and analyzing data for research and risk assessment purposes. Basically, the idea is to create a database to collect detailed data about all the women that underwent breast surgery with placement of breast implants: (1) specifying the type of surgery (cosmetic or reconstructive); (2) describing which brand and specifications of implant used in surgery; (3) identifying adverse problems after surgery (e.g. information about rupture, causes of inflammatory reaction and any other reactions). This database would allow detailed monitoring of possible problems that a breast implant brands could evidence. This implementation would have been of a tremendous help to patients, medical staff and health care planners for the early identification of quality problems of PIP breast implants. There is still a need for a better reporting on breast implant failures, in particular on ruptures, through the mandatory vigilance reporting system to identify potential design problems earlier [8].

The Institute of Medicine emphasizes the need for the physical and chemical characteristics of implants must be available, and standards imposed to ensure that future changes are made only when a research shows that no complications or other harmful effects on safety and health can occur [18].

In order to identify the problems inherent to breast implants, it is necessary to analyze the explanted implants. Until now, there are already a few tips on how the analyze the explanted implants should be carried out [9, 29]. Data collection is recommended at the time of explantation by a surgeon or appropriate healthcare provider at the explant site. A first procedure may be to collect information about the breast implants before proceeding to tests, to help in future comparison: implant type; lot number, volume, shell surface (smooth or textured), shape (round or anatomic), duration of implantation, implant placement (e.g., retromuscular, subglandular or subfacial), in vivo trauma (accident, mammography), among others. Another suggestion is to prepare a questionnaire about the patient's daily activities, such as sports practice or, any recent traumas. It would also be important if surgeons could identify the position of the implant in the chest cavity, and the rupture spot in the implant These data would be important to verify if the position and contact with the breast tissue influence the rupture, to evaluate if the mechanical behavior of tissues and internal forces during static postures and dynamic activities have any direct effects on the rupture. Surprisingly, there is no literature correlating mechanical loads on the breast tissues during daily or sport activities, and the main reactions between tissue and implant. Knowledge of the forces acting on the breast and its impact on the mechanical properties of the breast tissues is important for studying the design of breast implants as well as the effects of plastic surgery techniques for breast reconstruction [32].

Another information should be recorded, such as: reason(s) for the device explantation; presence of any shell defects; type of rupture (split, hole, v-shaped split or gross damaged); extent of implant rupture (intracapsular, extracapsular, or migrated gel); any discoloration, opacity of the shell and in the filler; and when the rupture occurred before or during explantation (if applicable). Figure 1 shows two examples of the possible characterization of the explanted implants, to describe the appearance of the shell, according to the Department of Health Therapeutic Goods Administration criteria [31].

The explanted implant should then be analyzed by mechanical testing, chemical analyses, and scanning electron microscopy (SEM). Microscopic (biopsy) of local tissue/capsule is also recommended. These analyses allow a categorization and documentation of the failure mode/mechanism, and are quite useful as a supplemental tool in understanding the mechanisms of implant failure [9].

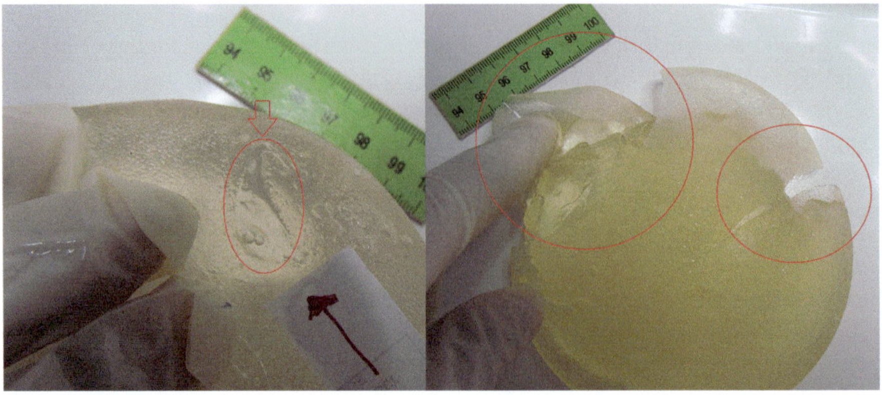

Fig. 1 Photographs of the two different types of shell rupture and color (left image: split rupture and clear shell; right image: gross damage and yellow color

According to the FDA [29] an analysis through microscopy techniques may provide details of the ruptured shell region and can be used to determine the cause of breast implant failure (e.g. instrument damage during surgery, fatigue, manufacturing defects, trauma), as published by Brandon et al. [9]. Further, research on the physical properties of breast implants shell is recommended, with the aim of finding an explanation for the rupture. The mechanical test should follow the standards that describe specific requirements for breast implants.

The impact of shell thickness should also be analyzed. A previous study, reported that differences in the shell thickness along the implant might affect the integrity of the silicone breast implant shell [18]. There are several recent studies that observed variability in thickness shell, where the minimum thickness was below 0.5 and maximum above 1.0 mm, despite the manufacturer's specifications ranged of 0.5 and 1.0 mm (FDA) [17–19, 22, 23, 29, 31].

The shell analysis can provide important information regarding its durability and the interaction with mechanical and chemical body properties. The manufacture of the shell involves curing of polymeric components of silicones by chemical crosslinking. During this process, the implant material, PDMS, starts with a range of molecular weights, and then a selected amount of "nano-particles" of amorphous "fumed" silica (SiO2) filler is added to liquid PDMS to make higher-performance silicone rubber (increases the strength and tear-resistance) [33]. The primary factor that controls the mechanical properties of filled silicone elastomers is the bonding between the silicone polymer and the filler (usually silica) [9]. If silica-silicone bonding (conventional cross-links between silicone molecules) are degraded the resistance of the material decreases. Suggested methods to determine the extent of crosslinking may include: the young modulus values, Fourier Transform Infrared Spectroscopy analysis, or measurement of equilibrium swelling (crosslink density).

In order to analyze the effects of implantation time on the durability of implant shells it is suggested to separate the implants according to type, so that explants can be compared with the proper controls. This is due to the fact that the strength of implant shells can vary considerably according to the manufacturer [9].

It is also recommended to identify the causes of irritant and other reactions (inflammation following rupture of an implant, capsular contracture…) between the breast implants and surrounding tissues. The PIP implants ruptures showed that most patients were asymptomatic by the time of explantation, although some symptomatic patients presented palpable lymph nodes in the axillae, change in the consistency and shape of the breast, development of a breast lump and, less commonly, pain. As intraoperative findings, augmented periprosthetic liquid, macroscopic presence of silicone particle on this liquid, stiffness and inflammatory changes on some capsules and axillary lymph nodes were observed as previously reported [34].

The Institute of Medicine committee advised a comparison of the stage of cancer detected by mammography in women with and without breast implant, and additional studies around the saline implants commonly used in reconstruction surgery [25].

Balk et al. [30] offered a fresh, forward-looking approach to the subject. They suggest two solutions to increase breast implant device surveillance, centered on maximizing the usefulness of the existing cohorts of women with breast implants. The first is the development of the National Breast Implant Registry (NBIR), that has been implemented since 2012, which collects and tracks important information on patients and devices. The American Society of Plastic Surgeons has committed to developing, funding, and operating this comprehensive NBIR in collaboration with the FDA. The second suggestion was to reanalyze the existing databases of large, ongoing clinical trials in the United States can be reanalyzed to provide more meaningful data on long-term systemic health effects. Integral to these solutions is ensuring that all pertinent patient information is collected and made available for directed analyses, particularly data on potential health confounders, such as family history, hormonal status, weight, and alcohol and tobacco use. By maximizing data extraction from existing clinical cohorts and directing more meaningful analyses, researchers may finally attain clarification of the long-term health outcomes of silicone gel breast implants.

Future work is needed to study some remaining gaps, for example (a) how leakage of small molecules can be minimized, (b) the mechanical robustness insured without compromising the performance of breast implants, and (c) how to improve the device-tissue boundary to minimize contraction risk. If the medical community and support laboratories address these issues, we would be closer to the development of future safer devices and improved safety standards. Figure 2 summarizes the recommendations for future work.

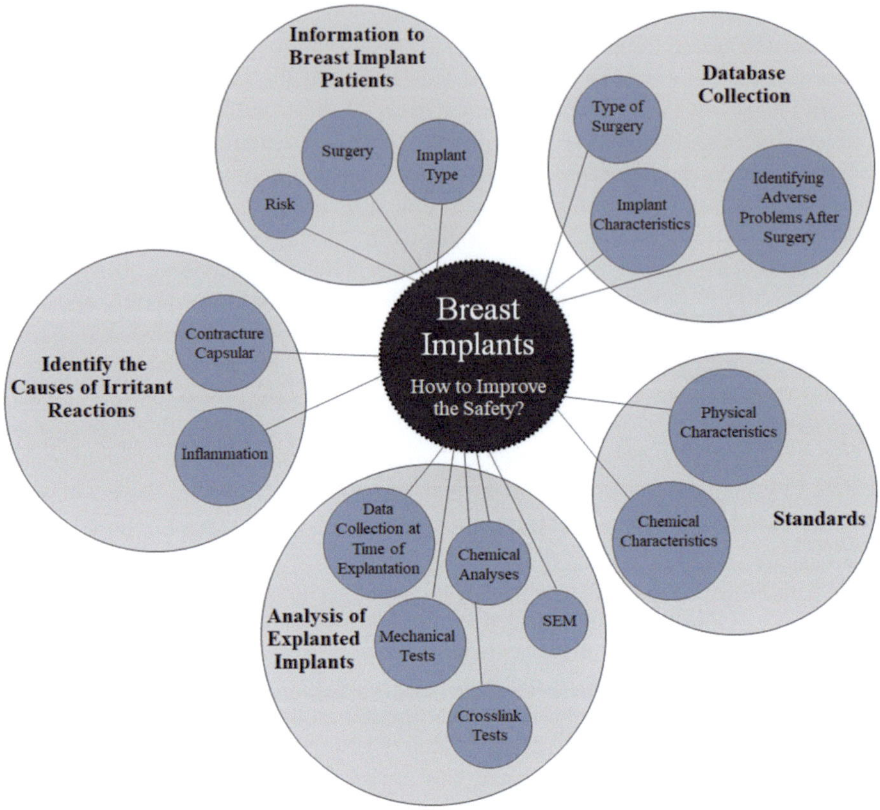

Fig. 2 Summary of the recommendations for future research

4 Conclusion

Inclusion of silicone breast prosthesis for aesthetic or reconstructive purposes offers temporary result of using a device to reach the new desired volume and in many occasions, revision surgeries are necessary due to internal and external factors that might affect the final result. The problems raised with the PIP implants, which apparently used non-medical degree silicone, show us that all the efforts to make the other brands safer have being successful making the devises safer and more reliable.

References

1. ISAPS (2014) International survey on aesthetic. Global Statistic. http://www.isaps.org/news/isaps-global-statistics. Accessed 11 Feb 2016
2. Maxwell GP, Gabriel A (2009) The evolution of breast implants. Clin Plast Surg 36(1):1–13. https://doi.org/10.1016/j.cps.2008.08.001
3. Bondurant S, Ernster V, Herdman R (2005) Safety of silicone breast implants. National Academies, Washington, pp 1–559
4. Food and Drug Administration (1992) (FDA), Special report. The basis of the FDA´s decision on breast implants. The New England Journal of Medicine-. New Engl J Med 326(25):1713–1715
5. Yu LT, La Torre G, Marotta J et al (1996) In vitro measurement of silicone bleed from breast implants. Plast Reconstr Surg 97:756–764
6. Siemionow MZ, Marita ME (2010) Plastic and Reconstructive Surgery. Springer, London. ISBN 978-1-84882-512-3
7. Phillips JW, de Camara DL, Lockwood MD et al (1996) Strength of silicone breast implants. Plast Reconstr Surg 97(6):1215–1225
8. Goldberg EP, Widenhouse C, Marotta J et al (1997) Failure of silicone gel breast implants: analysis of literature data for 1652 explanted prostheses. Plast Reconstr Surg 100:281–284
9. Brandon HJ, Jerina KL, Wolf CJ (2012) Retrieval and analysis of breast implants emphasizing strength, durability, and failure mechanisms. Woodhead Publishing Limited. https://doi.org/10.1016/B978-1-84569-799-0.50007-4
10. Cunningham B, (2007) The mentor core study on silicone memory gel breast implants. Plast Reconstr Surg 120(9S–29S):30–32. https://doi.org/10.1097/01.prs.0000286574.88752.04
11. Spear SL, Murphy DK, Slicton A et al (2007) Inamed silicone breast implant core study results at 6 years. Plast Reconstr Surg 120(8S–16S):17–18. https://doi.org/10.1097/01.prs.0000286580.93214
12. Barr S, Bayat A (2011) Breast implant surface development: perspectives on development and manufacture. Aesthet Surg J 31:56–67. https://doi.org/10.1177/1090820X10390921
13. Rohrich RJ, Adams WP, Beran SJ et al (1998) An analysis of silicone gel-filled breast implants: diagnosis and failure rates. Plast Reconstr Surg 102(7):2304–2308
14. Scientific Committee on Emerging and Newly Identified Health Risks (2013) The safety of PIP silicone breast implants. http://ec.europa.eu/health/scientific_committees/emerging/docs/scenihr_o_043.pdf. Accessed 10 Mar 2013
15. Wazir U, Kasem A, Mokbel K (2015) The clinical implications of poly implant prothèse breast implants: an overview. Arch Plast Surg 42:4–10. https://doi.org/10.5999/aps.2015.42.1.4
16. Santos DC, Barroso ML, Gomes N et al (2015) PIP breast implant rupture—A retrospective study from Portugal. Eur J Plast Surg 38(4):301–308
17. Swarts E, Kop AM, Nilasaroya A et al (2013) Rupture of poly implant prothèse silicone breast implants: an implant retrieval study. Plast Reconstr Surg 131(4):480e–489e. https://doi.org/10.1097/PRS.0b013e3182818a00
18. Grigg M, Bondurant S, Ernster VL et al (2000) Information for women about the safety of silicone breast implants. National Academy Press, Washington D.C. ISBN 0-309-52021-5
19. Grewal NS, Fisher J (2013) Why do patients seek revisionary breast surgery? Aesthet Surg J 33(2):237–244. https://doi.org/10.1177/1090820X12472693
20. Minami E, Koh IH, Ferreira JC et al (2006) The composition and behavior of capsules around smooth and textured breast implants in pigs. Plast Reconstr Surg 118(4):874–884. https://doi.org/10.1097/01.prs.0000240878.24213.b7
21. Liu X, Zhou L, Pan F et al (2015) Comparison of the post-operative incidence rate of capsular contracture among different breast implants: a cumulative meta-analysis. PLOS ONE 10(2): e0116071. https://doi.org/10.1371/journal.pone.0116071

22. Brody GS, Long JN, Talavera F (2016) Silicone Breast Implant Safety and Efficacy. Medscape. http://emedicine.medscape.com/article/1275451-overview#showall. Accessed 9 Feb 2016
23. Fisher J, Handel N (2014) Problems in breast surgery a repair manual. CRS Press, Taylor and Fraqncis Group. ISBN 13:978-1-4822-6093
24. Headon H, Kasem A, Mokbel K (2015) Capsular contracture after breast augmentation: an update for clinical practice. Arch Plast Surg 42(5):453–532. https://doi.org/10.5999/aps.2015.42.5.532
25. Arquero PS, Zanata FC, Ferreira LM et al (2015) Capsular weakness around breast implant: a non-recognized complication. World J Plast Surg 4(2):168–174
26. Roxo AC, Nahas FX, Salin R et al (2016) Volumetric evaluation of the mammary gland and pectoralis major muscle following subglandular and submuscular breast augmentation. Plast Reconstr Surg 137(1):62–69. https://doi.org/10.1097/PRS.0000000000001874
27. Potter EH, Rohrich R, Bolden K (2013) The role of silicone granulomas in recurrent capsular contracture: a review of the literature and an approach to management. Plast Reconstr Surg 131(6):888e–895e. https://doi.org/10.1097/PRS.0b013e31828bd642
28. Austad ED (2002) Breast implant-related silicone granulomas: the literature and the litigation. Plast Reconstr Surg 109:1724–1732
29. Food and Drug Administration (FDA) (2006) Guidance for industry and FDA Staff. Saline, Silicone Gel, and Alternative Breast Implants. U.S. Department of Health and Human Services.
30. Balk EM, Earley A, Avendano EA, Raman G (2016) Long-term health outcomes in women with silicone gel breast implants: a systematic review. Ann Intern Med 164(3):164–175. https://doi.org/10.7326/M15-1169
31. TGA (2013) PIP breast implants: Update on TGA testing of PIP breast implants. http://www.tga.gov.au/alert/pip-breast-implants-update-tga-testing-pip-breast-implants. Accessed 10 Mar 2013
32. Ramião NG, Martins PS, Rynkevic R et al (2016) Biomechanical properties of breast tissue, a state-of-the-art review. Biomech Model Mechanobiol 15(5):1307–1323. https://doi.org/10.1007/s10237-016-0763-8
33. Daniels AU (2012) Silicone breast implant materials. Swiss Med Wkly 142:w136
34. Keogh SB (2012) Poly Implant Prothese (PIP) breast implants: final report of the working group. Leeds, UK: Department of Health, NHS Medical Directorate; 2012. Report No. 17744

Part III
The Biomechanics of the Reproductive Period

Anthropometrics and Ergonomics in Pregnant Women

Henrique A. Almeida, Rita M. T. Ascenso and Eunice Oliveira

Abstract Pregnancy is a women life event that impacts her own and offspring's life; surprisingly, few systematic and scientific maternal information is available in the literature for anthropometric and ergonomic purposes. Regarding ergonomics, some authors state that guidelines should be design for pregnant women. These guidelines may consider anthropometric issues such as changes in weight and distribution of mass, which influence mobility, home and labour performance, mood and cognition. Anthropometrics allows the body measurement, but in pregnancy women are sparingly characterized, besides Body Mass Index and few other parameters. Ergonomics not only aids in the design process of work or home layouts, but also contains several evaluation tools to determine the level of risk of musculoskeletal disorders of daily or labour tasks. Some of these tools are very specific and others have certain conditions that should be met before using them. The purpose of this study is to present an overview of guidelines and discuss which anthropometric and ergonomic evaluation tools may be used, and under which conditions can be implemented in pregnant women.

Keywords Pregnancy · Anthropometrics · Somatotype · Ergonomics
Guidelines for pregnant women

H. A. Almeida (✉) · R. M. T. Ascenso · E. Oliveira
School of Technology and Management,
Polytechnic Institute of Leiria, Leiria, Portugal
e-mail: henrique.almeida@ipleiria.pt

R. M. T. Ascenso
Center for Research in Informatics and Communications,
Polytechnic Institute of Leiria, Leiria, Portugal

E. Oliveira
R&D Unit INESC Coimbra, Coimbra, Portugal

S. Brandão et al. (eds.), *Women's Health and Biomechanics*,
Lecture Notes in Computational Vision and Biomechanics 29,
https://doi.org/10.1007/978-3-319-71574-2_8

Highlights

- Anthropometric profile, characteristics and variations of pregnant women are presented in this chapter
- Risk assessment tools for musculoskeletal disorders of working pregnant women, ergonomic considerations and guidelines to accommodate pregnant women regarding home and labour are also addressed.

1 Introduction

Due to the professional and economic demand on working women, many of those forced to work late into their pregnancies and returning to work less than three months after giving birth. While much attention has been paid to the importance of ergonomics in the workplace, practicing good ergonomics will provide comfort and support the woman's health throughout her pregnancy. Decreasing the ergonomic risk factors by implementing reasonable accommodations are vital to make the workplace both safe and comfortable for expecting women. Work-related challenges can arise during the different stages of the pregnancy and after the woman has returned to work, depending on the individual and her working condition, meaning that it is important to adjust the tasks and workload. The basis for ergonomics is the morphological characterization that is fully obtained through anthropometrics. Also, anthropometric profile characterizes body composition closely related to health condition. Its relevance increases during pregnancy, not only because it has a direct impact on women's wellbeing but also impacts offspring. The aim of this chapter is to present an overview of both anthropometric characterization of pregnant women and ergonomic guidelines, and evaluation tools that may contribute to better choices in working places and wellbeing.

2 Anthropometric Profile During Pregnancy

There have been several studies that focus on body composition and anthropometrics. However, there are few metrics and standards suited for characterizing the body morphology and composition of pregnant women during pregnancy. Currently there are standard data for certain anthropometric parameters and specific groups such as children and seniors. Yet, specific data and equations for pregnant women have already been developed, but there's still need for more information to whether or not it is possible to use the existing data such as somatotype, body fat, Body Mass Index (BMI), Waist/Hip Ratio (WHR) and the Conicity Index (CI) [1].

Anthropometry allows the characterization of morphology and body composition through non-invasive methods, which determine the anthropometric profile. It may be a complete or restrict profile; whereas the last one includes the somatotype,

Table 1 Body Mass Index (BMI) classification

Category	Pregnant women[a]	Adults > 20yrs[b]
Underweight	<19.8	<18.5
Normal weight	19.8–26.0	18.5–24.9
Overweight	>26.0–29.0	25.0–29.9
Obese	>29.0	>30.0

[a]Institute of Medicine, 1990 [3]
[b]WHO—BMI for adults with more than 20 yrs old [4]

body fat, BMI, WHR and CI calculation, which evaluates waist circumference in relation to height and weight. Notwithstanding being a non-invasive method, this requires a total of 17 measurements.

In 1998, Paxton and colleagues developed two equations for the study of body fat to be used in pregnant women with different BMI, different gestational weight gains, different ethnicities and different socioeconomic status. This was based on the longitudinal study of body composition from weeks 14 to 37 of pregnancy in 200 black, white and hispanic women of New York between January 1991 and January 1994 [2].

The BMI classification is different for adult individuals and pregnant women (Table 1).

Some of the existing standards and parameters may be used regarding pregnant women, such as the CI, WHR and Abdominal-Hip Ratio (AHR), as the results presented normal values within the tabulated values for adults, but for other parameters was demonstrated the need to create new reference tables for anthropometric characterization of pregnant women, as demonstrated by Santana et al. [5]. This study considered 7 pregnant women between the ages of 26 and 40 with a time period of gestation between 21 and 35 weeks. Santana et al. also refer that a higher comprehensive study regarding both a longer gestation time period and the number of pregnant participants is needed [5].

Fluegel et al. [6] measured the changes from the 16th week of pregnancy on 198 german women revealing that the waist circumference increased by 27%, weight by 17%, chest circumference by 6%, and hip circumference by 4%. Similar increases have been reported by Rutter et al. [7] for 105 american women with an average age of 26 years. In average, the body weight increases 17%, and 8% in chest, 4% in hip, and 2% in abdominal circumferences at 16 weeks of pregnancy. Overweight and obese women show smaller increase in the abdominal region than those who were not [7].

Mid-Upper Arm Circumference (MUAC) is often used to assess nutrition levels during pregnancy, but no universal cut-off points have been identified. Various national nutritional protocols use the following MUAC cut-off values for inclusion of pregnant women into supplementary feeding programmes, namely in African countries. Malnutrition is considered for values of MUAC [8]:

- <18.5 cm in Zimbabwe, 2008;

- <21.0 cm in Burkina Faso and Burundi, 2002, Democratic Republic of Congo 2008, Guinea 2005, Madagascar 2007, Malawi 2007, Mali 2007, Niger 2006, Senegal, 2008;
- <22.0 cm in Mozambique, 2008;
- <22.5 cm in Zambia, 2009.

Some nutritional protocols enrol pregnant women based on gestational age (mostly only in the third trimester) regardless of any anthropometric measurement, but these studies revealed the relevance of MUAC [9]. There is no consensus on how to identify pregnant women as acutely malnourished and when to enrol them in nutritional programmes [8].

Several physical classification systems have been proposed over the decades, such as the somatotype, initially proposed by Sheldon (1940) and later on modified by others, namely Parnell (1958) and Heath and Carter (1967) [1]. The determination of the somatotype of an individual allows to evaluate the body shape and composition. Each individual is characterized according to three parameters (biotypes) assigned with a scale from 1 to 12 for each parameter. Endomorph is related to the relative fat (adiposity), mesomorph is related to the relative musculoskeletal robustness (muscularity) and ectomorph is related to the linearity of the body (thinness). The assigned number indicates the magnitude of each parameter. The most common values stay between 1 and 7; from 1 to 2.5 are considered very low for each biotype, while very high values are considered above 7.5. Each parameter is influenced by the individual's stature. Only with the combination of three biotypes, endomorph + mesomorph + ectomorph (in this order) in a unique sort expression defines the somatotype, which are then converted to a somatopoint in order to mark it in a graph called somatochart [1]. Besides this tool being the most complete anthropometric evaluation tool, neither the authors of the tool or other researchers have presented conclusive somatotype studies of pregnant women.

In 1974 at Tartu University, a body structure study included 678 non-pregnant women (18 to 22 years old) and 3919 pregnant women (18 to 29 years old). It is a well-known fact that the course of pregnancy and parturition and the health of the new-born are closely connected with the individual dimensions of the mother's body such as height, weight, external and internal pelvic measurements, which influence the course of labour [10].

The amount of healthy weight gain during a pregnancy varies [11]. Weight gain is only partly related to the weight of the baby and growing placenta, and includes extra fluid for circulation, and the weight needed to provide nutrition for the growing *foetus*. Most needed weight gain occurs later in pregnancy [12]. During pregnancy, insufficient or excessive weight gain can compromise the health of the mother and *foetus*. The most effective interventions for weight gain in underweight women are not clear [12]. Being or becoming very overweight in pregnancy increases the risk of complications for both the mother and the *foetus* including a caesarean birth, gestational hypertension and diabetes, pre-eclampsia, macrosomia and shoulder dystocia [11]. Besides these complications, the losing the excess weight is more difficult after pregnancy in such cases [11, 13]. The American

Table 2 Average pregnancy weight gain according to the women's body mass index [14, 15]

BMI before pregnancy	BMI category	Recommended weight gain during pregnancy (kg)
less than 18.5	Underweight	12.7–18.0
18.5–24.9	Normal weight	11.3–15.9
25.0–29.9	Overweight	6.8–11.3
above 30.0	Obese	5.0–9.0

College of Obstetricians and Gynaecologists [14] and the Institute of Medicine [3] proposed the average pregnancy weight gain according to the women's BMI before pregnancy (refer to Table 2).

Changes in anthropometric and body composition characteristics during pregnancy was studied in 406 healthy, pregnant women aged between 16 and 33 years old in West Bengal, India, revealing that the mean change in maternal weight from the first to the third trimester was merely 3 kg. Mean waist circumference varied from the first to the third trimester but not from the first to the second trimester [16].

The influence of weight occurs during the whole reproductive cycle. *Postpartum* weight retention at 6 months and weight gain from 6 to 18 months *postpartum* seems to have equal contribution adverse maternal anthropometric measures 7 years after delivery [17]. Pregnancy changes the women's body during pregnancy and the following months or even years after, influencing not only the daily routines but also their working place. In some cases, even before pregnancy, some women need to undergo a decrease or increase in weight in order to improve the success of getting pregnant.

Besides the increase in body mass, its distribution also differs which impacts the mother's centre of mass. This topic is very critical when designing products that are influenced by this shift of the centre of mass. The increasing abdominal protrusion of pregnant women also makes it increasingly difficult to get as close to working objects as they could before. The working area of the hands becomes smaller during pregnancy, and manipulating objects that are now further ahead of the spinal column will generate an increased compression and bending strain on the spine, ligaments and muscles in the back. This loading is caused by the increasing mass of the abdomen and its increasing moment arm with respect to the spinal column. Endresen [18] reported that the occupational factor with the greatest risk for pelvic pain and/or lower back pain is twisting or bending several times an hour.

3 Home and Labour Performance

As mentioned before, with the changes in body dimensions, the abdomen becomes increasingly larger, causing progressive postural problems, backache, and diminishing of dexterity, agility, coordination and balance. In consequence of that, physical performance capabilities at home and work are reduced. The working area of the hands (mobility) and distant object manipulation (far reaches) tends to decrease even more if an exertion of large energy is need, and particularly if repetitive or over long periods of time [19].

Nicholls and Grieve [20] surveyed 200 pregnant (between 29 and 33 weeks pregnant) living in London. Women were asked about their current performance on certain tasks compared with that from before pregnancy. The interviews used a five-point ordinal scale and concerned 46 activities preselected to be included in the study. Out of these 46 tasks, 32 were found to be significantly more difficult to perform during pregnancy. Among these, the following were considered the hardest: picking up an object from the floor; working at a desk; walking upstairs; driving a car; getting in and out of a car; and using seat belts; ironing; reaching high shelves; using public toilets; and getting in and out of bed. The reasons reported for the difficulties were related to back pain, reduced reach and clearance, feeling unstable, being fatigued, having reduced mobility, and having difficulties in seeing objects near the body.

Several biological, physiological, social, biomechanical, occupational factors and body-shape changes that occur during pregnancy contribute to the development of back pain at work [21–25]. Pregnant women sitting 8-10 h a day at a desk or working in a computer refer back pain and neck injuries, stiff muscles and joints, poor circulation and worker's fatigue [26, 27]. In a survey of 72 pregnant women, Cheng et al. [28] found that sitting was the most frequent problematic task at work because they got tired, felt an uncomfortable posture and spent excessive time in the same posture [29].

Dumas et al. [29] evaluated the effect of improving a conventional desk by adding a concave "desk board" designed to provide arm support during computer work on the back and upper extremity for women in late pregnancy (Fig. 1). This study revealed that during computer work, pregnant women sat with a more upright posture than non-pregnant. The muscle activity of the anterior deltoid was higher

Fig. 1 **a** Pregnant participant completing the 20 min computer task using the standard desk setup in the laboratory. **b** WorkplaceTM board set up on top of the standard desk in the laboratory. **c** Pregnant participant completing the 20 min computer task using the WorkplaceTM board in the laboratory [29]

for the pregnant than for the control group. The pregnant also reported more discomfort in the lower back and pelvis area. The ergonomic board tested in the study did not affect trunk or neck posture, but decreased muscle activity in some back muscles. Its effect on the upper extremity is less desirable as it promoted wrist extension, elbow flexion and shoulder abduction, as well as increased muscle activity in the extensor *digitorum* and right trapezius.

As pregnancy progresses, women must interact with objects at distances further away from their body, having to reach between 40 to 50 cm away. Since the growing *foetus* alters the centre of gravity, women are also more susceptible to falling. In addition to affecting balance, lifting tasks, posture, and previous back pain cases are associated with increased risk of developing low back pain during pregnancy, which suggests a direct link between ergonomic stressors and unfavourable pregnancy results [21, 30–33]. Women should avoid standing for long periods of time, working long hours, and performing repetitive lifting [19]. In summary, the following work-related strains can affect both the *foetus* and/or the pregnant woman's health:

- Hard physical labour and heavy lifting;
- Activities involving standing and/or walking for long periods of the workday;
- Long working hours with few opportunities for breaks;
- Seated work over long periods of time;
- Constricted workplaces (can cause unfavourable working positions);
- Stress;
- Vibrations.

4 Risk Assessment Tools for Musculoskeletal Disorders of Working Pregnant Women

In ergonomics, there are several risk assessment tools for musculoskeletal disorders, but only a couple mention or propose risk assessment tools for pregnant women: Manual Handling Assessment Charts (MAC) and the National Institute for Occupational Safety and Health (NIOSH).

Regarding the MAC, it contains the following recommendation "The vulnerability of special risk groups (e.g. pregnant women, young workers etc.) should be considered where appropriate." In this particular scenario, as one uses the MAC worksheet, one must choose the options accordingly to the pregnant women. However, no particular multiplier or additional factor is considered in the worksheet.

Regarding the NIOSH, the previous NIOSH Revised Lifting Equation was adapted to derive recommended weight limits for pregnant workers and to develop

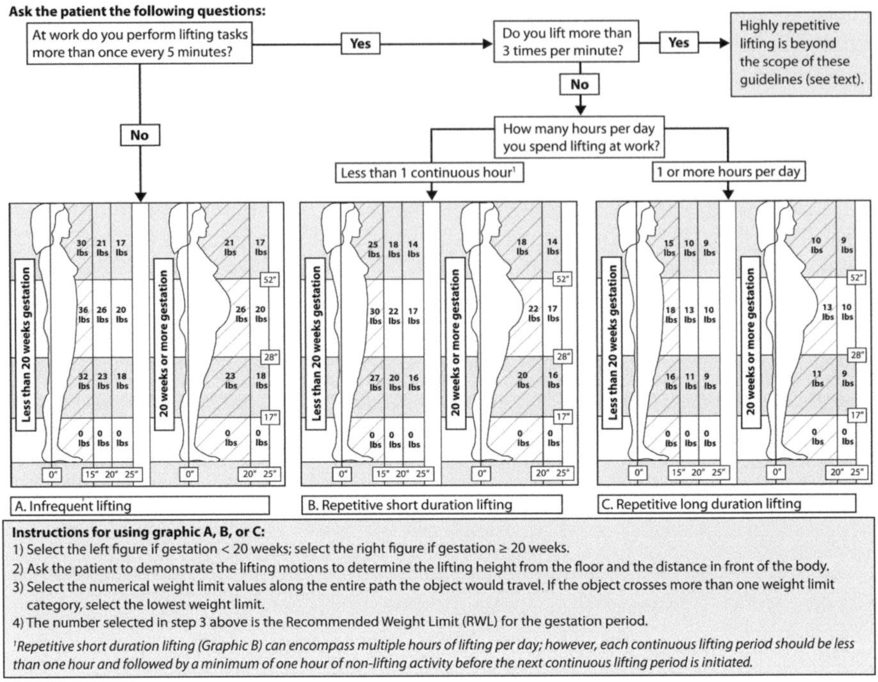

Fig. 2 Recommended weight limits in early and late pregnancy for three lift frequency patterns [35]

corresponding guidelines for clinicians [34]. Waters et al. [34] determined lifting thresholds that most pregnant workers with uncomplicated pregnancies should be able to perform without increased risk of adverse maternal and *foetal* health consequences. The only exceptions are for restrictions involving lifting from the floor and overhead. Figure 2 presents the NIOSH risk assessment tool for pregnant workers.

5 Guidelines to Accommodate Pregnant Women

To accommodate pregnant women, either at the workplace, in vehicles, or at home, the following ergonomic considerations should be taken into account [19, 25, 32, 34, 36, 37]:

- The belly is growing, meaning that the sitting position must be adjusted, therefore, easy adjustable seats must be provided. Investing in adjustable comfortable ergonomic office chairs with an excellent back lumbar support and well cushioned seat supporting both the growing body and possible sensitive tailbone, or adding a thin lumbar pillow for extra support and comfort is a critical aspect during pregnancy.

- During gestation, many body changing dimensions occur, in particular the increasing abdominal protrusion. This makes it difficult to reach distant objects, making it necessary to adjust the height of working surfaces. In order to accommodate the growing abdomen and the increasing curve in one's spine, the height of the monitor and desk must also be adjusted. In summary, areas of manipulation should be kept close to the spinal axis of the body and elevated from their regular height.
- Not only do pregnant women loss reaching capability but they are also more susceptible to falling, therefore, the amount of work that is performed at heights must be reduced.
- Daily comfort only results from a neutral posture. In order to have a neutral posture, the elbow's level should at the same level of the keyboard so that the shoulders are relaxed and the feet should be flat and supported with a footrest. By doing so, awkward postures are avoided. The use of footrests or an anti-fatigue mats reduces swelling and pressure on the articular joints.
- In order to avoid static postures, pregnant women should vary their tasks. A varied working position aids in the reduction of strain on both the back and pelvis, and also ensures a good circulation throughout the body.
- Standing should be limited to less than three hours per day.
- Frequent breaks from sitting, such as stretching out or walking around for a while must be implemented. This allows to increase the blood flow in the legs and feet, which are likely to swelling and possible blood clots during pregnancy.
- As mentioned before, variations between sitting and standing increases both oxygen intake and blood flow. Tables with height adjustment enables the pregnant employee to vary her working position between sitting, "sitting-to-stand", and standing along the workday. This position modification is also a mild form of exercise which prevents stiffness, swelling, and varicose veins.
- In order to prevent the risk of developing the carpal tunnel syndrome which is caused by increased fluid in the joints during pregnancy, a softer wrist rest with one's keyboard is very important.
- It is important to adjust the workload and tasks to the pregnant avoiding hard, stressful or uncomfortable tasks.
- Less physical tasks must be assigned to pregnant women. In the vertical direction, the work tasks should be reduced to minimum as possible. Object lifting must be avoided and safe lifting techniques must be promoted. Lifting items directly from ground level or heavy lifting above 5–10 kg according to one's physical capability must be avoided.
- The work must be adjusted during pregnancy (flexible scheduling, day shift rather than night shifts, etc.). Specific times of day when the employee is in a better condition and/or a workday that allows the pregnant employee to take more frequent and short breaks must be chosen.
- Regarding the workspace, more space than usual for moving around should be provided, and obstacles, in particularly low objects that might be difficult to see, must be avoided.

- The need for more frequent breaks for eating and drinking during the workday might be necessary.
- Both the employer and employee must know their rights. Both should be able to follow the legislation concerning the treatment of pregnant employees, and the pregnant women should not hesitate in requiring for special treatment, according to her physical condition.

6 Conclusions

Pregnancy is one of the most common life events and therefore several issues must be considered for supplying healthy conditions during pregnancy. A better understanding of the body morphology and composition will give improved insights to the body modifications during pregnancy. In consideration to existing anthropometric evaluation tools, further research is needed. Regarding ergonomics, some authors state that design guidelines should be provided for pregnant women, including issues such as changes in size, distribution of mass, mobility, performance, mood and cognition. Ergonomics not only aids in the design process of work or home layouts, but also contains several evaluation tools to determine the level of risk of musculoskeletal disorders of daily or labour tasks. Some of these tools are very specific while others have to be further investigated and validated before using them.

Acknowledgements The authors acknowledge the support of a Project Grant (UID/MULTI/00308/2013) funded by the Portuguese Foundation for Science and Technology.

References

1. Norton K, Olds T (2005) Antropométrica. ARTMED editora
2. Paxton A, Lederman SA, Heymsfield SB et al (1998) Anthropometric equations for studying body fat in pregnant women. Am J Clin Nutr 67(1):104–110
3. Institute of Medicine (1990) Nutrition During Pregnancy: Part I: Weight Gain, Part II: Nutrient Supplements. The National Academies Press, Washington, DC, p 480
4. WHO http://apps.who.int/bmi/index.jsp?introPage=intro_3.html Accessed June 2015
5. Santana MT, Pereira RJM (2015) Perfil Antropométrico Restrito De Grávidas. Biomechanichos Degree Research Project, School of Technology and Management, Polytechnic Institute of Leiria, Leiria, Portugal
6. Fluegel B, Griel H, Sommer K (1986) Anthropologischer Atlas. Tribuene, Berlin, Germany
7. Rutter GB, Haager AJ, Daigle GC, Smith S, Mcfarland N, Kelsey N (1984) Dimensional changes throughout pregnancy: a preliminary report. Carle Select Pap Fall 36(2):38–46
8. Ververs M-t, Antierens A, Sackl A, Staderini N, Captier V (2013) Which anthropometric indicators identify a pregnant woman as acutely malnourished and predict adverse birth outcomes in the humanitarian context? PLoS Currents 5: ecurrents.dis.54a58b618c611bc031ea140e613f2934599c293459

9. Kerac M, McGrath M, Seal A (2010) Management of acute malnutrition in infants (MAMI) Project. Technical review: current evidence, policies, practices & programme outcomes. IASC, ACF, UCL, Emergency Nutrition Network
10. Kaarma H, Kasmel J, Peterson J, Veldre G (2001) Application of nonpregnant and pregnant women's anthropometric data in medicine. Mank Q 42
11. Viswanathan M, Siega-Riz AM, Moos MK, Deierlein A, Mumford S, Knaack J, Thieda P, Lux LJ, Lohr KN (2008) Outcomes of maternal weight gain. Evid Rep Technol Assess 168
12. IQEHC—Institute for Quality and Efficiency in Health Care (2014) Weight gain in pregnancy fact sheet. Institute for Quality and Efficiency in Health Care
13. Thangaratinam S, Rogozińska E, Jolly K et al (2012) Interventions to reduce or prevent obesity in pregnant women: a systematic review. Health Technol Assess. 16(31):iii–iv, 1–91. https://doi.org/10.3310/hta16310
14. ACOG—American College of Obstetricians and Gynecologists (2013) ACOG Committee opinion no. 548: weight gain during. Obstet Gynecol. 121(1):210–212. https://doi.org/10.1097/01.AOG.0000425668.87506.4c
15. Institute of Medicine (2009) Weight gain during pregnancy: Re-examinging the guidelines, Report Brief
16. Ghosh A (2012) Anthropometric and body composition characteristics during pregnancy: a study from West Bengal, India. Homo 63(3):233–240. https://doi.org/10.1016/j.jchb.2012.03.003
17. Kirkegaard H, Stovring H, Rasmussen KM et al (2014) How do pregnancy-related weight changes and breastfeeding relate to maternal weight and BMI-adjusted waist circumference after delivery? Results from a path analysis. Am J Clin Nutr 99(2):312–319. https://doi.org/10.3945/ajcn.113.067405
18. Endresen EH (1995) Pelvic pain and low back pain in pregnant women—an epidemiological study. Scand J Rheumatol 24(3):135–141
19. Kroemer KHE, Kroemer HB, Kroemer-Elbert KE (2000) Ergonomics: How to design for ease and efficiency. Prentice-Hall
20. Nicholls JA, Grieve DW (1992) Perform Phys Tasks Pregnancy Ergonomics 35(3):301–311
21. Cheng PL, Pantel M, Smith JT et al (2009) Back pain of working pregnant women: identification of associated occupational factors. Appl Ergon 40(3):419–423. https://doi.org/10.1016/j.apergo.2008.11.002
22. Levangie PK (1999) Association of low back pain with self-reported risk factors among patients seeking physical therapy services. Phys Ther 79(8):757–766
23. MacEvilly M, Buggy D (1996) Back pain and pregnancy: a review. Pain 64(3):405–414
24. Orvieto E, Achiron A, Ben-Rafae Z et al (1994) Low-back pain of pregnancy. Acta Obstet Gynecol Scand 73(3):209–214
25. Paul JA, van Dijk FJ, Frings-Dresen MH (1994) Work load and musculoskeletal complaints during pregnancy. Scand J Work Environ Health 20(3):153–159
26. Hartvigsen J, Leboeuf-Yde C, Lings S et al (2000) Is sitting-while-at-work associated with low back pain? A systematic, critical literature review. Scand J Public Health. 28:230–239
27. Dumas GA, Reid JG, Wolfe LA et al (1995) Exercise, posture and back pain during pregnancy. Part II: Exercise and back pain. Clin Biomech 10(2):104–109
28. Cheng PL, Dumas GA, Smith JT et al (2006) Analysis of self-reported problematic tasks for pregnant women. Ergonomics 49(3):282–292. https://doi.org/10.1080/00140130500434929
29. Dumas G, Upjohn T, Delisle A et al (2009) Posture and muscle activity of pregnant women during computer work and effect of an ergonomic desk board attachment. Int J Ind Ergon 39:313–325. https://doi.org/10.1016/j.ergon.2008.03.003
30. Rungee JL (1993) Low back pain during pregnancy. Orthopedics 16(12):1339–1344
31. Cherry N (1987) Physical demands of work and health complaints among women working late in pregnancy. Ergonomics 30(4):689–701
32. Heliövaara M (1989) Risk factors for low back pain and sciatica. Ann Med 21(4):257–264
33. Fung BK, Kwong CM, Ho ES (1993) Low back pain of women during pregnancy in the mountainous district of central Taiwan. Zhonghua Yi Xue Za Zhi (Taipei) 51(1):103–106

34. Waters TR, MacDonald LA, Hudock SD et al (2014) Provisional recommended weight limits for manual lifting during pregnancy. Hum Factors J Hum Factors Ergon Soc 56(1):203–214. https://doi.org/10.1177/0018720813502223
35. MacDonald LA, Waters TR, Napolitano PG et al (2013) Clinical guidelines for occupational lifting in pregnancy: evidence summary and provisional recommendations. Am J Obstet Gynecol 209(2):80–88. https://doi.org/10.1016/j.ajog.2013.02.047
36. Ireland ML, Ott SM (2000) The effects of pregnancy on the musculoskeletal system. Clin Orthop Relat Res 372:169–179
37. Morrissey SJ (1998) Work place design recommendations for the pregnant worker. Int J Ind Ergon 21(5):383–395. https://doi.org/10.1016/S0169-8141(96)00079-0

Increased Step Width During Walking as Pregnancy Progresses: Functional or Mechanical Adaptation?

Wendy L. Gilleard

Abstract Women change their gait as pregnancy progresses. It is known biomechanical adaptions occur; however, the etiology of these adaptations is unknown. A common suggestion is that the adaptations, such as increased step width, are a functional response to control stability. Nevertheless, step width is also affected by trunk segment kinematics which are themselves also altered as pregnancy progresses. Step width is also affected by mechanical obstruction when the girth of the thigh is increased. Therefore, the changes seen in step width may be of mechanical rather than functional origins. The objective of this chapter is to investigate the etiology of the step width adaptation as pregnancy progresses and its resolution post-birth. It includes discussion on the temporospatial and trunk segment kinematic adaptions as pregnancy progresses and the effect of parity. The mechanical obstruction by increased thigh girth is a likely explanation for increased step width in late pregnancy. The mechanical effect of altered trunk mechanics may also affect step width as pregnancy progresses. Thus, increased step width as pregnancy progresses is a mechanical adaptation rather than a functional adaptation to increase stability. The effect from previous pregnancy may be less important than the individual's differences in adaptations.

Keywords Pregnancy · Post-birth · Parity · Step width · Walking
Adaptations

W. L. Gilleard (✉)
School of Health and Human Sciences, Southern Cross University, Lismore, NSW, Australia
e-mail: Wendy.Gilleard@scu.edu.au

S. Brandão et al. (eds.), *Women's Health and Biomechanics*,
Lecture Notes in Computational Vision and Biomechanics 29,
https://doi.org/10.1007/978-3-319-71574-2_9

Highlights

- Step width increase as pregnancy progress is due to increased thigh girth
- The effect of previous pregnancy may be less important than the individual's differences in adaptations.

1 Introduction

It has long been recognized that women change their gait as pregnancy progresses with anecdotal comments of a waddling gait. There is some consensus in the literature as to the biomechanical adaptations that occur. However, the etiology of these adaptations is not well understood. A common suggestion in the literature is that the adaptations, such as increased step width as pregnancy progresses, are a functional response to control stability. Instability and falling is a common problem in pregnancy reported as early as 1943 [1]. Step width effects, or are affected by, trunk segment kinematics during walking [2], which themselves are also changed as pregnancy progresses [3]. Step width is also altered by mechanical obstruction when the girth of the thigh is increased [4]. Therefore, the changes seen in step width may be of mechanical rather than functional origins.

The purpose of this chapter is to discuss and investigate the etiology of the step width adaptation as pregnancy progresses and its resolution post-birth.

2 Biomechanical Adaptions When Walking During Pregnancy

Walking is an essential daily activity and important in controlling adipose tissue gain associated with pregnancy [5]. As pregnancy progresses with consequential increased mass and segment girths [6], adaptations are reported in the biomechanics of walking. Adaptions as pregnancy progresses are seen in both temporospatial and kinematic characteristics.

While several studies have investigated temporospatial characteristics of gait in pregnancy the results are equivocal. Walking velocity has been reported be slower in late pregnancy [7–10], although this is disputed [3, 11–14]. Stride length is reported to decrease as pregnancy progresses [3, 9, 13, 14] although no significant change has been reported [11, 15].

Step length is reported as not changing significantly [8, 12, 15] or decreased in late pregnancy [10, 13, 14]. Cadence (frequency of strides per minute) is reported to be significantly decreased [10] or remain unchanged in late pregnancy [11, 13]. There are several reports of a significant increase in single and double support [9–11, 14], although others report no significant change in double support time [8, 13]. Step width is commonly reported as increased in late pregnancy [3, 8, 10, 11, 13,

15] although Branco et al. (2013) [14] reports no significant change. Comparisons between studies are difficult due to differences in study design, weeks of gestation when testing occurred, and often the lack of a nulliparous control group.

The etiology of temporospatial gait adaptions is not well understood. Several authors suggest that the adaptations seen when pregnant women are walking may be a mechanism to control stability [3, 8–10, 13, 14]. Characteristics such as velocity, stride length and step width effect, or are affected by, trunk segment kinematics during walking [2, 16].

As pregnancy progresses, the lower trunk segment inertial characteristics show a significantly larger rate of increase than any other body segments [6]. With rapid changes in mass, and moment of inertia [6], trunk segment kinematics are altered in daily activities such as walking, similar to that seen when carrying loads on the trunk [17]. In the sagittal plane, Foti et al. (2000) [11] reported increased peak anterior pelvic tilt in late pregnancy when compared to one year post-birth, despite no significant changes to the pelvic segment range of motion [3, 11, 18]. Gilleard (2013) [3] reported reduced sagittal plane range of motion for the thoracic segment in late pregnancy, supporting McCrory et al. (2014) [18] who reported the range of motion was reduced in late pregnancy compared to the second trimester and the control group.

The range of motion in the transverse plane of the pelvic [11, 18, 19] and thoracic segments [3, 18, 19], are reported to be not significantly altered in late pregnancy. In contrast, Gilleard (2013) [3] reported significant linear trend for decreased range of motion as pregnancy progressed in the pelvic segment. Interestingly while, Wu et al. (2010) [7] found that the range of motion amplitudes were similar to a control nulliparous group, the intra-individual standard deviations were significantly smaller. The authors included women between 20 and 34 weeks-gestation in a single group for which there would have been wide variation in the lower trunk segment inertias. The within group variability may have precluded finding a significant difference in amplitude although a consistent reduction in comparison to the control group was noted. Gilleard (2013) [3] reported a significant linear trend for decrease in the range of motion of the pelvic segment in the coronal plane (side tilt) in contrast to Foti et al. (2000) [11] and McCory et al. (2014) [18] who reported no change in late pregnancy. There is no significant change in the range of motion for the thoracic segment [3, 18].

2.1 Importance of Longitudinal Study Design and Use of Nulliparous Control Group

Until recently, much of the existing literature on biomechanical adaptions in pregnancy was based on designs where participants tested in late pregnancy act as their own controls, and are again tested post-birth. In this design approach, testing post-birth is implicitly assumed to reflect the non-pregnant musculoskeletal state.

While this approach has increased our knowledge of biomechanical adaptations for some parameters, we now understand that there may be some continuing effects of pregnancy or the time period for resolving an adaption may be some time after birth. Therefore, the use of a post-birth test as a participant control has limited validity in relation to the effects of pregnancy.

Cross sectional designs, where discrete control groups have been used, typically report the control group as non pregnant females in a similar age group at times matched for height and non pregnant mass. Often no report is given as to whether they are nulliparous (a woman who has never carried a pregnancy beyond 20 weeks) or indeed they may be multiparous. If we are to understand the changes related to adaptations to pregnancy it is important that the control group is nulliparous. Therefore, we assume that all data from the control group reflects a true non-pregnant state with no potential ongoing effects of a pregnancy beyond 20 weeks.

While understanding of the effects of pregnancy on trunk segment motion during walking and temporospatial characteristics has increased using cross sectional or comparison to post-birth designs, further information is required about adaptations as pregnancy progresses and in the early post-birth period. Pregnancy is characterized by changes in dimensions and mass that occur over a typical period of 40 weeks. As such it is important to understand that changes may occur at any time over the pregnancy and may not all be at their maximum adaption in late pregnancy. A longitudinal repeated measures design over pregnancy and post-birth, with a matched nulliparous control group would provide insight into the mechanical adaptations of the body under conditions of changing load, while minimizing the influences of inter-human differences. Hence, in order to fully understand adaptions related to pregnancy, testing should be undertaken at a minimum in each trimester using a longitudinal design.

Study design has impacted on the understanding of mechanical adaptions as pregnancy progresses. Simple designs such as use of the post-birth test as a participant self-control or cross sectional designs, avoid the issue of poor availability of participants willing to be tested multiple times in a laboratory setting which requires changes to their normal daily routine. However, studies where a longitudinal retest approach have been used, have shown that the adaptions seen are often individual and the use of statistical approach based on group results may mask the inter-individual differences. While prospective studies that include a baseline test prior to pregnancy are highly desirable and participants are then able to act as their own controls, such designs are extremely difficult to conduct and there are only few reports [10].

2.2 Post-birth Adaptions

At parturition there is an immediate reduction in the mass of the maternal trunk segment. Despite this rapid change in mass, the structural adaptions required to accommodate the growing fetus during the 40 weeks of pregnancy take time to readapt. Gilleard and Brown (1996) [20] reported continuing functional deficits in the abdominal muscles, in parallel with incomplete resolution of structural adaptations at 8 weeks post-birth [20]. Interestingly in a later study Gilleard (2013) [3] reported altered sagittal range of motion in the pelvic and thoracic segments at 8-weeks post-birth which was not seen during pregnancy. This adaption seen only post-birth may be related to the functional capability of anterior abdominal wall musculature and the posterior trunk muscles at this time. Stabilization of the pelvis by the abdominal muscles is compromised up to 8 weeks post-birth [20] at the same time the fatigability of the trunk extensors is decreased [21]. Thus, an imbalance between the anterior and posterior postural muscles may exist post-birth, and this may be reflected in the decreased pelvic segment range of motion. The increased range of motion of the thoracic segment may reflect a counter motion to minimize the load on the thoracolumbar spine [3]. There is a paucity of reports on trunk segment biomechanics in the later post-birth period. Therefore, it is unknown whether these adaptions remained unresolved.

In order to be able to investigate post-birth adaption, comparison must be made to either a control group or pre pregnancy data. Forczek and Staszkiewicz (2012) [10] using a prospective design, found that all temporospatial biomechanical adaption seen in late pregnancy were reversed by six months post birth [10] and were not significantly different to the results of pre pregnancy testing. Interestingly Bertuit et al. (2015) [13] reported that while step width was resolved the velocity, step and stride length remained decreased, and the time in double support remained increased [13]. While the average time post-birth was 6 months, the group ranged from 4 to 8 months post-birth [13]. Therefore, it is possible that the variation in time for re-adaptation across the group confounded the results.

2.3 *Potential Effect of Parity*

While temporospatial adaptations in the biomechanics of walking are thought to be resolved by 6 months post-birth, the pattern of re-adaptation for the changes seen in the mechanics of the trunk segment as pregnancy progresses are unknown. It is possible that the biomechanical effects of pregnancy may continue and therefore influence the biomechanics of walking for subsequent pregnancies. Dumas et al. (1995) [22] also reported an effect of parity on standing posture. Regardless of whether they exercised regularly, multiparous women had a significantly larger lumbar lordosis curvature than primiparous women [22]. It is possible that the imbalance between the anterior and posterior postural muscles seen at eight weeks post birth [20] may continue. Gilleard et al. (2002) [23] reported that adaptions of the pelvic segment sagittal plane angle continued into the post-birth period for at least 8 weeks post-birth for quiet standing [23]. It is also possible that habitual postures adopted in the first pregnancy continue [23]. Habitual motion patterns may also continue.

3 Biomechanics of Step Width Adaptations as Pregnancy Progresses

Most of the literature reports step width is increased in late pregnancy [3, 10, 11, 13, 15, 18] and the increase is resolved post birth [3, 8, 10, 13]. While it is commonly suggested that increases in step width are a mechanism to control stability, there is a paucity of research investigating the etiology of increased step width as pregnancy progresses.

3.1 Etiology of Increased Step Width: Mechanics to Increase Stability, Mechanical Obstruction of the Thigh or Effect of Altered Kinematics of the Trunk?

It is commonly reported that biomechanical adaptations seen in walking during pregnancy are a mechanism to control stability [3, 8–10, 13, 14]. It also possible that the biomechanical adaptions are a mechanical consequence related to increased mass and segment girths resulting in mechanical obstruction to segment motion and or an alteration in inertial control required due to the change in mass [8, 13]. However, the causes of the increased step width seen during pregnancy are unknown. An understanding of the etiology of increased step width during a normal pregnancy will inform the current body of knowledge of the musculoskeletal demands of a normal pregnancy.

Step width is affected by biomechanical parameters which themselves also change during pregnancy. Step width affects, or is affected by, trunk segment kinematics during walking in both older and younger adults [2]. Variations in frontal plane trunk kinematics effects the step-by-step variations in the consequent step width with 54% of the variance in step width being accounted for by frontal plane trunk horizontal position (28%) and acceleration (15%) respectively [2]. Proprioception of the horizontal position in the frontal plane of the trunk determines the subsequent step width required to maintain the center of mass within the base of support [2]. During late pregnancy the trunk horizontal motion side to side is increased as pregnancy progresses [18]. Therefore it is possible that trunk frontal plane angular motion may influence the step width as pregnancy progresses.

Step width is also affected by thigh girth [4]. Increased step width is seen in obesity [24] where, similar to pregnancy, an increase in mass and dimensions occurs. Increased step width was found when the thigh circumference was increased, both with and without additional mass [4]. Hence, step width is affected primarily by mechanical obstruction rather than an adaption for inertial control.

While it is known that the dimensions of the thigh increase as pregnancy progresses [6], there is a scarcity of studies investigating the effect of thigh girth on step width in late pregnancy.

3.2 *Current Research on the Etiology of Step Width Adaptions as Pregnancy Progresses*

Step width is known to increase as pregnancy progresses and is thought to be a mechanism to control stability [3, 10, 13]. However, step width relates to trunk segment kinematics during walking [2], and is also influenced by thigh girth [4]. This means that it is possible that the step width adaptions are a mechanical consequence related to increased thigh segment girths resulting in obstruction to segment motion. It is also possible that altered trunk frontal plane kinematics, in an attempt to maintain inertial control of the trunk may also affect the step width. Gilleard and Bradbury [25] investigated the factors affecting step width as pregnancy progresses and the re-adaption in the post-birth period. The study investigated the relationship between step width and thigh girth, the frontal plane range of motion (ROM) of the trunk, and the frontal plane trunk ROM relative to pelvis, at 38 weeks gestation, at 8 weeks post-birth and in a nulliparous control group [25].

Eight pregnant women were investigated when walking along a 20-meter walkway at a self-selected velocity using an 8-camera motion analysis system at 38 weeks gestation and again at eight weeks post birth. A control group of 11 non-pregnant nulliparous women were tested once. Markers defining the thoracic and pelvic segments and also on each lateral malleolus were included. Thigh girth was recorded at each test session at the gluteal fold. Data was collected for three trials per test session and the data taken from the third trial used in the subsequent analysis. Walking velocity was determined for each trial. Linear regression analysis was used to investigate the effect of dependent variables on the variance of step width at each test. Results showed no significant difference in walking velocity between the groups ($p = 0.89$). At 38 weeks gestation, variation in step width was influenced by thigh girth with 67% ($p = 0.01$) of the step width variance explained (Table 1). Supporting this result, significant increases in step width were reported by Westlake et al. (2013) [4] when thigh girth was increased by an average of 1.3 cm. Thus, mechanical obstruction caused by increased thigh circumference is a likely reason for increased step width in late pregnancy. This influence was not seen post-birth nor in the nulliparous control group. At 38 weeks gestation by the frontal plane motion of the trunk relative to the pelvis also helped to explain the variation with 46% ($p = 0.06$) of the step width variance explained. The effect was not seen post-birth or in the control group. The frontal plane motion of the trunk relative to the pelvis is decreased in late pregnancy only [3] and was mainly influenced by reduced frontal plane motion of the pelvis. Further research is warranted to explore the effect of frontal plane motion of the pelvis on the variance in step width in late

Table 1 Regression analysis results for the independent variable step width and the dependent variable of Frontal Plane Trunk ROM, and Frontal trunk ROM relative to Pelvis

Variable r^2 (p, t)	38 weeks	Post birth	Controls
Thigh girth	0.67 (0.01, 3.54)*	0.00 (0.96, 0.04)	0.00 (0.82, 0.23)
Frontal plane trunk ROM relative to pelvis	0.46 (0.06, 2.27)*	0.11 (0.41, 0.89)	0.18 (0.19, −1.42)
Frontal plane trunk ROM	0.22 (0.24, −1.30)	0.23 (0.22, −1.35)	0.33 (0.06, −2.15)*

*Significant at $p < 0.1$

pregnancy. Between 22–33% of the variance in step width was accounted for by frontal plane trunk ROM for each group (Table 1), similar to the results for trunk center of mass position from Hurt et al. (2010) [2]. As frontal plane trunk ROM is not changed significantly as pregnancy progresses [3], a similar influence on step width would be expected to be seen at post birth and also for nulliparous controls.

There is a strong relationship between thigh girth and step width in late pregnancy that is not seen post-birth nor in controls. These results indicate the mechanical obstruction by increased thigh girth is a likely explanation for increased step width in late pregnancy rather than increased step width being a functional adaptation to increase stability.

4 Effect of Multiple Pregnancies on Step Width

Temporospatial adaptations in the biomechanics of walking, such as step width, are thought to be resolved by around 6 months post-birth. It is possible however that the adaptations as pregnancy progresses may vary with parity, with a previous pregnancy influencing the response to subsequent pregnancies. In order to more fully understand the effects of pregnancy, a longitudinal data collection consisting of data prior to pregnancy (nulliparous), a first pregnancy (primiparous) and a second pregnancy (multiparous) would be required.

Gilleard (2015) [26] investigated the effect of a first and subsequent pregnancy on step width when walking [26]. One nulliparous subject was initially tested twice (Nulliparous). Four test sessions over each pregnancy and one post-birth were conducted for two ensuing pregnancies (Pregnancy 1 and 2). Thigh girth at the gluteal fold was measured using a flexible tape at each test session. An 8-camera motion analysis system recorded walking at a self-selected velocity and the results were graphed. For comparative purposes, Maternal and Control group data from a previous study [3] were included. Results were based on visual interpretation of the graphs.

Thigh girth increased as each pregnancy progressed. Thigh girth at 38 weeks in Pregnancies 1 and 2 were 68 cm and 61 cm, respectively. Nulliparous step width

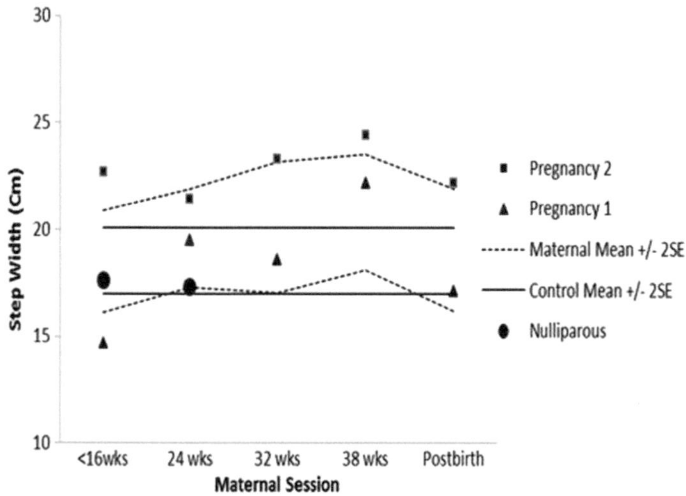

Fig. 1 Step width for Nulliparous, Pregnancy 1 and 2. Maternal and Control group data from Gilleard (2013) [3]

seen in Fig. 1, was similar to the Controls. For Pregnancy 1 and 2 the step width increased as pregnancy progressed and decreased post birth in a similar pattern to the Maternal group. These changes are in agreement with the literature [3, 10, 13]. Therefore, the mechanical adaptation to pregnancy was similar in pattern for each pregnancy.

There appeared to be an effect of parity as step width was consistently larger in Pregnancy 2 and was also larger than the one from Nulliparous. The etiology of this increased step width for Pregnancy 2 is unclear. The thigh girth although increased over both pregnancies was smaller in Pregnancy 2 than Pregnancy 1. Therefore, although there was a mechanical influence of the increased thigh girth on step width, it is possible that the increase itself rather than the magnitude of the increase is the influencing factor. It is also possible that there were habitual walking patterns already established prior to Pregnancy 2 related to increased loads from carrying a small child. Further investigation is warranted to investigate the effect of step width when walking for women who have children aged less than 2 yrs. The results also indicated that a woman's mechanical response to each pregnancy and resolution post birth differs. The carry over effect from previous pregnancy may be less important than the differences in adaptations by an individual.

5 Conclusions

The mechanical obstruction by increased thigh girth is a likely explanation for increased step width in late pregnancy. The mechanical effect of altered trunk mechanics may also affect step width as pregnancy progresses. Increased step width as pregnancy progresses is a mechanical adaptation rather than a functional adaptation to increase stability. There are also many other influences on step width, such a habitual motion patterns and life factors such as carrying small children in subsequent pregnancies. The response to each pregnancy is also distinct. Therefore, the effect from previous pregnancy may be less important than the individual's differences in adaptations.

References

1. Fries E, Hellebrandt FA (1943) The influence of pregnancy on the location of the centre of gravity, postural stability and body alignment. Am J Obstet Gynecol 46(3):374–380
2. Hurt CP, Rosenblatt N, Crenshaw JR et al (2010) Variation in trunk kinematics influences variation in step width during treadmill walking by older and younger adults. Gait Posture 31 (4):461–464. https://doi.org/10.1016/j.gaitpost.2010.02.001
3. Gilleard WL (2013) Trunk motion and gait characteristics of pregnant women when walking: report of a longitudinal study with a control group. BMC Pregnancy Childbirth 13:71. https://doi.org/10.1186/1471-2393-13-71
4. Westlake CG, Milner CE, Zhang S et al (2013) Do thigh circumference and mass changes alter knee biomechanics during walking? Gait Posture 37(3):359–362. https://doi.org/10.1016/j.gaitpost.2012.07.031
5. Maturi MS, Afshary P, Abedi P (2011) Effect of physical activity intervention based on a pedometer on physical activity level and anthropometric measures after childbirth: a randomized controlled trial. BMC Pregnancy Childbirth 11:103. https://doi.org/10.1186/1471-2393-11-103
6. Jensen RK, Doucet S, Treitz T (1996) Changes in segment mass and mass distribution during pregnancy. J Biomech 29(2):251–256
7. Wu W, Meijer OG, Lamoth CJ et al (2004) Gait coordination in pregnancy: transverse pelvic and thoracic rotations and their relative phase. Clin Biomech 19:480–488. https://doi.org/10.1016/j.clinbiomech.2004.02.003
8. Lymbery J, Gilleard W (2005) The stance phase of walking during late pregnancy. J Am Podiatr Med Assoc 95(3):247–253
9. Błaszczyk JW, Opala-Berdzik A, Plewa M (2016) Adaptive changes in spatiotemporal gait characteristics in women during pregnancy. Gait Posture 43:160–164. https://doi.org/10.1016/j.gaitpost.2015.09.016
10. Forczek W, Staszkiewicz R (2012) Changes of kinematic gait parameters due to pregnancy. Acta Bioeng Biomech 14(4):113–119
11. Foti T, Davids J, Bagley A (2000) A biomechanical analysis of gait during pregnancy. J Bone Joint Surg Am 82(5):625–662
12. Golomer E, Ducher D, Arfi GS et al (1991) Etude de la locomotion simple et lors du transport de charge chez la femme enceinte. J Gynecol Obstet Biol Reprod (Paris) 20(3):406–412
13. Bertuit J, Feipel V, Rooze M (2015) Temporal and spatial parameters of gait during pregnancy. Acta Bioeng Biomech 17(2):93–101

14. Branco M, Santos-Rocha R, Aguiar L et al (2013) Kinematic analysis of gait in the second and third trimesters of pregnancy. J Pregnancy 2013:718095. https://doi.org/10.1155/2013/718095
15. Bird AR, Menz HB, Hyde CC (1999) The effect of pregnancy on footprint parameters: a prospective investigation. J Am Podiatr Med Assoc 89(8):405–409
16. Huang Y, Meijer OG, Lin J et al (2010) The effects of stride length and stride frequency on trunk coordination in human walking. Gait Posture 31(4):444–449. https://doi.org/10.1016/j.gaitpost.2010.01.019
17. LaFiandra M, Wagenaar RC, Holt KG et al (2003) How do load carriage and walking speed influence trunk coordination and stride parameters? J Biomech 36:87–95
18. McCrory JL, Chambers AJ, Daftary A et al (2014) The pregnant "waddle": an evaluation of torso kinematics in pregnancy. J Biomech 47(12):2964–2968. https://doi.org/10.1016/j.jbiomech.2014.07.009
19. Wu WH, Huag Z-J, Meiger OG et al (2010) Motor coordination mode of transverse pelvic and thoracic rotations during walking in healthy pregnant women. J Clin Rehabil Tissue Eng Res 14(2):375–380
20. Gilleard WL, Brown HMM (1996) Structure and function of the abdominal muscles in primigravid subjects during pregnancy and the immediate postbirth period. Phys Ther 76 (7):750–762
21. Dumas G, Adams M (1001) Dolan P (1998) Fatigability of back extensors after pregnancy. J Biomech 1998(31 Suppl):33
22. Dumas GA, Reid JG, Wolfe LA et al (1995) Exercise, posture, and back pain during pregnancy: Part 1. Exerc Posture Clin Biomech 10(2):98–103
23. Gilleard W, Crosbie J, Smith R (2002) Static trunk posture in sitting and standing during pregnancy and early post-birth. Arch Phys Med Rehabil 83:1739–1744. https://doi.org/10.1053/apmr.2002.36069
24. Spyropoulos P, Pisciotta JC, Pavlou KN et al (1991) Biomechanical gait analysis in obese men. Arch Phys Med Rehabil 72:1065–1070
25. Gilleard W, Bradbury J (2016) Increased step width during pregnancy when walking is related to increased thigh girth. In: ABC10 2016, Melbourne
26. Gilleard W. (2015) Effect of parity on step width during walking as pregnancy progresses: a case study. In: BioMedWomen—international conference on clinical and bioengineering for women's health 2015, Porto, Portugal

Diastasis Recti During Pregnancy and *Postpartum*

Patrícia Mota, Augusto Gil Pascoal, Cristiana Vaz, Filipa João, António Veloso and Kari Bø

Abstract *Diastasis recti abdominis* (DRA) or increased *inter-rectus* distance (IRD) is characterized by the separation of the rectus abdominis muscles. It has its onset during pregnancy and the first weeks following childbirth. The reliability of the instruments used to assess this condition is unclear. There is scant knowledge on the prevalence and risk factors for development of the condition. There is little evidence on which exercises are most effective in reduction of DRA. The aims of our studies were to establish a reliable method for the assessment of the morphology of the abdominal wall, describe DRA prevalence, risk factors, and evaluate the acute response on the IRD induced by drawing-in and abdominal crunch exercises. The results of three methodological studies showed ultrasound imaging to be a reliable method for measuring IRD. The ultrasound transducer can be held relatively stationary in a clinical setting, to evaluate IRD. DRA is prevalent at 6 months postpartum, with a prevalence rate of 39%. The acute response on IRD produced by drawing-in exercise was a widening of the IRD in postpartum, while the abdominal crunch exercise induced an acute narrowing response of the IRD in pregnancy and in postpartum.

Keywords *Diastasis recti* · *Inter-rectus* distance · Ultrasound · *Postpartum*

P. Mota (✉) · A. G. Pascoal · C. Vaz · F. João · A. Veloso
Fac. Motricidade Humana, CIPER, LBMF, Univ Lisboa, Lisbon, Portugal
e-mail: patimota@gmail.com

P. Mota
ESTeSL, Escola Superior de Tecnologia da Saúde de Lisboa,
Instituto Politécnico de Lisboa, Lisbon, Portugal

K. Bø
Department of Sports Medicine, Norwegian School of Sports Sciences, Oslo, Norway

S. Brandão et al. (eds.), *Women's Health and Biomechanics*,
Lecture Notes in Computational Vision and Biomechanics 29,
https://doi.org/10.1007/978-3-319-71574-2_10

Highlights

- Ultrasound Imaging is a reliable method to assess *inter-rectus* distance (IRD), and the ultrasound transducer can be held relatively stationary in a clinical setting, to evaluate IRD during abdominal crunch and drawing-in exercises
- The increased IRD or *diastasis recti* (DRA) is prevalent at 6 months postpartum but not linked with lumbo-pelvic pain
- During pregnancy and postpartum period IRD narrows during abdominal crunch, and widens during drawing-in exercise.

1 Introduction

Pregnancy and motherhood are exciting times in a woman's life. Besides all the hormonal and physiological changes affecting women during this period, probably the most obvious morphological alteration during pregnancy is the increasing weight and dimensions of the uterus, influencing maternal trunk musculoskeletal morphology, particularly the abdominal musculature. Many women continue or even begin to exercise during pregnancy, and shortly after delivery women are encouraged to resume abdominal exercises to restore their abdominal figure and fitness. A lot of information is available about exercise programs for women during pregnancy and in the *postpartum* period, and physiotherapists and exercise instructors prescribe exercises to this population everyday. However, there is little evidence available about muscular changes and the effect and safety of different abdominal exercises during and after pregnancy.

Diastasis recti (DRA) or the increased *inter-rectus* distance (IRD) seems to be a common condition in women during pregnancy and *postpartum.* The lack of evidence for the consequences of this condition and the effect of abdominal strengthening exercises in the reduction of DRA indicates a need for identification of prevalence and risk factors of DRA. Use of responsive, reliable, and valid outcome measures is mandatory to evaluate this condition, and ultrasound imaging has recently been suggested as a useful method to assess muscular geometry and to quantify the DRA.

2 Changes in the Abdominal Wall Morphology During Pregnancy

The functional role of the abdominal muscles during pregnancy appears to be similar to those in the nonpregnant state [1], suggesting that it is important for the movement of the trunk, pelvic stabilization, and restraint of the abdominal contents [2]. However, the musculoskeletal morphology of the anterolateral wall of the abdomen changes as pregnancy progresses [3, 4]. The weight and dimensions of the uterus and its contents increases from 40 to 1000 g, and its capacity from 4 ml in the nonpregnant state to 4000 ml at term [5]. The maternal inferior thoracic diameter is increased [6, 7] as well as the anterior and lateral dimensions of the abdomen. These changes modify the spatial relationship between the superior and the inferior abdominal muscle attachments [5] increasing the length of the abdominal muscles, particularly the *rectus abdominis* [8]. At 38 weeks of gestation, the length of the abdominal muscles increases around 115% with respect to the beginning of pregnancy [3]. The increment of the anterior abdominal dimensions may alter the angle of the abdominal muscle attachment in the sagittal plane [3]. Changes in the spatial relationship of muscle attachment and muscle's angle of insertion may alter the muscle's line of action and therefore their ability to produce torque [3, 7].

3 *Inter-rectus* Distance and *Diastasis Recti Abdominis*

One of the muscles thought to undergo adjustments during pregnancy is the *rectus abdominis*. As the fetus grows, the two muscle bellies of the *rectus abdominis*, connected by the *linea alba*, elongate, and curve round as the abdominal wall expands, with most separation occurring at the umbilicus [1, 3, 8]. The augmented IRD, often referred as DRA, is described as a change in the abdominal musculature, specifically in the *linea alba* and *rectus abdominis* sheath, with onset in the last trimester of pregnancy and whose peak of incidence occurs immediately after birth and the first weeks following childbirth [1, 9–11]. Although some studies suggested that an augmented IRD could reduce the abdominal integrity and functional strength, contributing to pelvic instability and back pain [3, 7, 12], no scientific evidence exists about the functional implications of an augmented IRD or even about the effect of the exercise on prevention and/or reduction of IRD.

4 Classification and Prevalence of *Diastasis Recti Abdominis*

Criteria and IRD cut-off value for the diagnosis of DRA vary in the literature [1, 3, 13–19], and to date, there is no international *consensus* on where to perform measurement. In a cadaver study, Rath et al. (1996) defined a widening of the IRD more than 10 mm above, 9 mm below and 27 mm at the level of the *umbilicus*, as pathological DRA. Others defined DRA as a widening of the IRD more than 25 mm at one or more assessment points using digital calipers [20]. In a more recent ultrasound study, Beer et al. (2009) suggest that, in nulliparous women, the *linea alba* should be considered "normal" when the IRD width is less than 15 mm, at the xiphoid level, 22 mm at 3 cm above, and 16 mm at 2 cm below the *umbilicus*.

Studies have found that DRA may affect between 30 and 70% of pregnant women [1], and that it may remain separated in the immediate *postpartum* period in 35 to 60% of women [13]. In a longitudinal study following a cohort with ultrasound assessment of the IRD from late pregnancy till 6 months *postpartum*, 39% of the women were diagnosed with DRA [21]. However, the condition has also been found in 39% of older, parous women undergoing abdominal hysterectomy [14] and in 52% of urogynecological menopausal patients [17]. The prevalence of DRA or increased IRD varies in the literature, also due to different measurement assessment methods and IRD cut-off values applied for diagnosis [1, 3, 13, 15, 17, 18, 20].

5 Risk Factors for *Diastasis Recti Abdominis*

There is scant knowledge about the risk factors for DRA. Two studies analyzed several variables such as age, ethnicity, body mass index, height, weight gain during pregnancy, pre-pregnancy weight, gestational age at delivery, type, and duration of birth [10, 16]. An association of DRA during pregnancy with Caucasian ethnicity and lack of regular exercise during pregnancy was suggested [16]. It is considered that women with DRA have a greater number of pregnancies and deliveries [10, 17], and among multiparous women, it is suggested that there is a strong association between provision of childcare and DRA during pregnancy, related to frequent lifting and carrying of small children which may increase strain on the abdominal wall and increase loading of the already weakened abdominal muscles during pregnancy [16]. Recently no risk factors were identified for the presence of DRA in one longitudinal study. However, these studies were limited by the sample size, reliability of the instruments used, and were not definitive in its ability to delineate risk factors.

6 The Effect of Exercise on *Diastasis Recti Abdominis*

It has been suggested that *antepartum* activity level may have a protective effect on DRA and exercise may improve *postpartum* symptoms of DRA [19]. Postnatal women are encouraged to resume abdominal exercises shortly after delivery to restore their abdominal figure and fitness [3, 6]. To date, there is scarce knowledge on the most effective abdominal exercises both during pregnancy and after childbirth. In particular, there is little evidence on which exercises are most effective in reduction of the *diastasis recti*. The *rationale* behind abdominal strengthening programs is the assumption that the contraction of all abdominal muscles reduces the horizontal diameter of the abdomen in such a way that a horizontal force will be generated producing the approximation of both *rectus abdominis*, particularly at the *umbilicus* level [22]. However, there is no evidence that this horizontal tension will produce an approximation of the *rectus abdominis* muscles. The horizontal force is the result of the overall action of the deep abdominal muscles (oblique's and *transversus abdominis* muscles) which are anteriorly attached to the lateral side of each *rectus abdominis* muscles [23] and posteriorly connected to the lumbar vertebral column. Thus, the horizontal tension produced by these deep abdominal muscles could pull the *rectus abdominis* muscle laterally toward the fixed sites on the vertebral column, increasing the IRD [24].

The abdominal crunch is one of the most used exercises in abdominal strengthening programs. However, the abdominal crunch has been considered a risk exercise for development of DRA [1], and lately, core training with the drawing-in exercise has been recommended both in the general population [25–27] and during pregnancy, and after childbirth [28]. It has been proposed that the activation and training of the *transversus abdominis* draw the bellies of the *rectus abdominis* muscle together [29, 30], improves the integrity of the *linea alba* and increases fascial tension, allowing efficient load transference and torque production [28, 31]. However, due to the low number and quality of the studies, there is insufficient evidence to support this statement. Additionally, in one longitudinal descriptive exploratory study on drawing-in (DI) and abdominal crunch during pregnancy and postpartum [32], there was a contrasting effect of the two exercises, with the abdominal crunch exercise consistently producing a significant narrowing of the IRD. In contrast, the DI exercise generally led to small widening of the IRD.

7 Ultrasound Imaging Transducer Orientation and Displacement Procedures and Instruments to Assess the *Inter-rectus* Distance

The most common methods to assess IRD are palpation [1, 13, 33, 34] and calipers [35, 36]. However, the reported prevalence of DRA (or augmented IRD) may be inaccurate because of the lack of reliability, low responsiveness (ability of a tool to detect small differences or small changes) and lack of validity [37] of both methods and instruments used to measure the IRD. Recently, ultrasound imaging has been suggested as a useful method to assess muscular geometry and as an indirect measure of muscle activation via changes in muscle thickness during contraction [38, 39]. Coldron et al. (2008) used ultrasound to characterize *rectus abdominis* changes during the first year postpartum [11] and Mendes et al. (2007) claimed ultrasonography to be an accurate method to measure *diastasis recti* when compared with surgical compass during abdominoplasty [40].

Factors such as relocation of the original imaging site, reproduction of the same transducer pressure and orientation, as well as maintenance of a relatively stationary transducer position during muscle contraction, could adversely affect reliability [41] and accurate interpretation of ultrasound imaging and lead to erroneous conclusions [42–44].

8 Procedure for Transducer Motion Analysis

No information exists in the literature about the best alignment for ultrasound transducer during IRD measurement. Nevertheless, a recent systematic review [45] provides some information about measurement of muscle fascicle lengths and pennation angles that could be extrapolated for IRD measurements. Benard et al. [46] showed that muscle fascicle lengths and pennation angles could be underestimated or overestimated if ultrasound probe is not aligned with the plane of muscle fascicles. To determine this plane, two methods may be applied: 1. positioning the probe perpendicular to the skin [46, 47] or 2. adjusting the probe alignment until image quality is optimized [46–48]. This second method is also used for IRD measurements. It is assumed that when an optimized image is reached, the probe should aligned with the plane of both *rectus abdominis* muscles.

Additionally, the dynamic nature of ultrasound imaging studies undertaken during abdominal exercises may compromise the interpretation of data. As the position and inward pressure of an ultrasound transducer influence both the location and the shape of a structure of interest on the image, it is logical to assume that accurate measurement and interpretation will be influenced by changes in its orientation to the body surface [44]. Recent studies suggest that a range of transducer motion between 5° to 10° for angular displacement around the three axes of rotation and 10 mm on one plane of translation (inward/outward transducer motion), may be acceptable without image distorting and/or introducing measurement error [42, 43]. Keeping these guidelines for acceptable amounts of transducer motion in mind, it is prudent to determine the amount of ultrasound transducer motion that occurs during ultrasound imaging studies undertaken during the assessment of IRD under contraction conditions.

As transducer motion interferes with accurate measurement, it is interesting to be able to document the amount of ultrasound transducer motion (orientation and displacement) when ultrasound images are collected for IRD measurements, during static positions of the abdominal crunch and drawing-in exercises. This way, in one of our studies, motion capture was collected with a 13-camera Qualisys system (model: Oqus-300plus) operating at a frequency rate of 50 Hz. The transducer motion was tracked by means of a customized cluster composed by four non-collinear reflective markers fixed to the long axis of the transducer and by two additional markers virtually built at the lower extremity of the transducer using a digitizing pointer (Fig. 1).

Fig. 1 Ultrasound transducer with a customized cluster composed by four noncollinear reflective markers fixed to the long axis of the transducer

Pelvis position was tracked by means of four reflective markers applied to both anterior–superior iliac spines and iliac crests.

The Visual 3D software (Visual 3D Basic RT, C-Motion, Inc., Germantown, MD) was used for tridimensional reconstruction of the pelvis and the ultrasonography transducer in the resting position (static calibration). The tridimensional reconstruction of the model enabled us to quantify the position and rotation of the transducer relative to the pelvis segment. Two local reference coordinate systems were defined: 1. the pelvis reference coordinate system origin was in the midpoint between the right and left anterior iliac spines, and 2. the transducer reference system origin was located at the midpoint of the two virtual markers on the base of the transducer (Fig. 2).

We quantified the orientation of the transducer relative to the pelvis around the x-axis (cranial/caudal tilt of the transducer), y-axis (medial/lateral tilt of the transducer) and z-axis (clockwise/counterclockwise rotation of the transducer). The translational movement of the transducer along the yz plane was calculated through

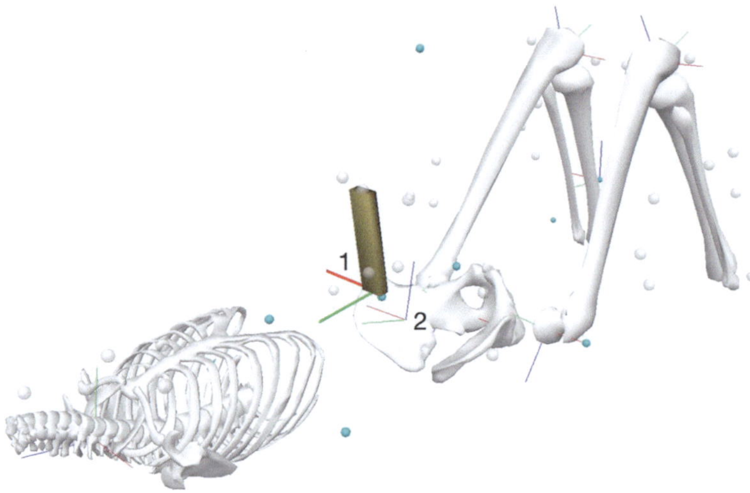

Fig. 2 Representation of the two local reference coordinate systems were: **1** the pelvis reference coordinate system,the origin of which was on the midpoint between the right and left anterior iliac spines; **2** transducer reference system the origin of which was located at the midpoint of the two virtual markers on the base of the transducer

the scalar distance of the midpoint between the two proximal markers placed in the transducer and the midpoint between the right and left anterior–superior iliac spines. The position of the transducer relative to the pelvis during the calibration of the subject (static motion capture) was assumed as zero degrees in all the axes of rotation and zero meters of translation regarding the linear position of the transducer.

None of the exercises (abdominal crunch and DI) produced large transducer motions relative to the pelvis, and all findings are within previously established guidelines for acceptable amounts of transducer motion (<10° of angular and 8 mm of inward/outward motion), except the translation during abdominal crunch exercise. Nevertheless, the highest mean value for translation during abdominal crunch in our study is <15 mm.

Besides the amount of transducer motion, our group decided to perform a test-retest reliability study to assess the intra-observer reliability of 2D ultrasound measurement of the distance between *rectus abdominis*, the IRD [49].

Ultrasound images from the *rectus abdominis* were recorded on 24 healthy female volunteers at rest and on two conditions of abdominal contraction: abdominal crunch and drawing-in exercises (Fig. 3). A blinded investigator measured the IRD offline from two different ultrasound images collected on two different days. Additionally, reanalyses of the same ultrasound images were performed on two separate occasions. Test-retest measurements of IRD demonstrated good to

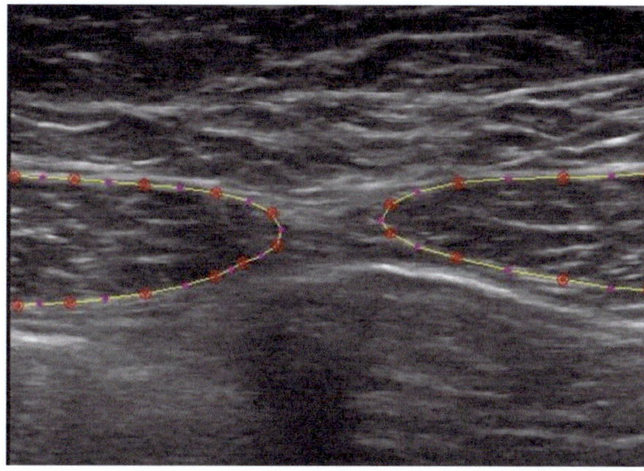

Fig. 3 *Rectus Abdominis* ultrasound image during rest. Points digitalized by the examiner on the muscles contour (red dots)

very good reliability with ICC values between 0.74 and 0.90. The only exception was for IRD measured 2 cm below the *umbilicus* on the abdominal crunch exercise, with an ICC of 0.50. For intra-tester reliability of the same images, the ICC values were all above 0.90 [49].

9 Conclusions and Future Directions

The dynamic nature of ultrasound imaging studies undertaken during abdominal exercises may compromise the interpretation of data, and it is logical to assume that accurate measurement and interpretation will be influenced by changes in the orientation of the transducer [44]. Nevertheless, our group found ultrasound imaging to be a reliable method for measuring the IRD at rest and during abdominal crunch and drawing-in exercises. Although ultrasound imaging is a reliable method to assess IRD, data on inter-rater reliability is needed when studies are conducted including more than one investigator.

The IRD cut-off value for categorizing DRA needs to be further studied. It could be interesting to study IRD values for women during pregnancy and postpartum longitudinally in a large sample size.

Further studies are needed to evaluate the effect of different abdominal exercises in the reduction of the IRD during the postpartum period. Given the high prevalence and the concern many women experience with this condition, further high quality randomized controlled trials on the effect of different abdominal exercises on the DRA are warranted.

References

1. Boissonnault JS, Blaschak MJ (1988) Incidence of diastasis recti abdominis during the childbearing year. Phys Ther 68(7):1082–1086
2. Standring S (2008) Gray's anatomy [Internet]. Elsevier Health Sciences
3. Gilleard WL, Brown JM (1996) Structure and function of the abdominal muscles in primigravid subjects during pregnancy and the immediate postbirth period. Phys Ther 76 (7):750–762
4. Foti T, Davids JR, Bagley A (2000) A biomechanical analysis of gait during pregnancy. J Bone Joint Surg Am 82(5):625–632
5. Cunningham F, Leveno K, Bloom S, Spong CY (2009) Williams obstetrics, 23rd edn. McGraw-Hill
6. Davies GAL, Wolfe LA, Mottola MF et al (2003) Joint SOGC/CSEP clinical practice guideline: exercise in pregnancy and the postpartum period. Can J Appl Physiol 28(3):330–341
7. Gilleard W, Crosbie J, Smith R (2002) Effect of pregnancy on trunk range of motion when sitting and standing. Acta Obstet Gynecol Scand 81(11):1011–1020
8. Fast A, Weiss L, Ducommun EJ et al (1990) Low-back pain in pregnancy. Abdominal muscles, sit-up performance, and back pain. Spine 15(1):28–30
9. Dumas GA, Reid JG, Wolfe LA et al (1995) Exercise, posture, and back pain during pregnancy. Clin Biomech 10(2):104–109
10. Rett MT, Braga MD, Bernardes NO et al (2009) Prevalence of diastasis of the *rectus abdominis* muscles immediately *postpartum*: comparison between primiparae and multiparae. Braz J Phys Ther 1 13(4):275–280. https://doi.org/10.1590/S1413-35552009005000037
11. Coldron Y, Stokes MJ, Newham DJ et al (2008) *Postpartum* characteristics of *rectus abdominis* on ultrasound imaging. Man Ther 13(2):112–121. https://doi.org/10.1016/j.math.2006.10.001
12. Parker MA, Millar LA, Dugan SA (2010) *Diastasis Rectus Abdominis* and lumbo-pelvic pain and dysfunction—are they related? J Womens Health Phys Ther. 32:1
13. Bursch SG (1987) Interrater reliability of diastasis recti abdominis measurement. Phys Ther 67(7):1077–1079
14. Ranney B (1990) *Diastasis recti* and umbilical hernia causes, recognition and repair. S D J Med 43(10):5–8
15. Rath AM, Attali P, Dumas JL et al (1996) The abdominal linea alba: an anatomo-radiologic and biomechanical study. Surg Radiol Anat 18(4):281–288
16. Candido G, Lo T, Janssen PA (2005) Risk factors for *diastasis of the recti abdomini*s. J Assoc Chart Physiother Womens Health 97:49–54
17. Spitznagle TM, Leong FC, Van Dillen LR (2007) Prevalence of *diastasis recti abdominis* in a urogynecological patient population. Int Urogynecol J Pelvic Floor Dysfunct 18(3):321–328. https://doi.org/10.1007/s00192-006-0143-5
18. Beer GM, Schuster A, Seifert B et al (2009) The normal width of the *linea alba* in nulliparous women. Clin Anat 22(6):706–711
19. Akram J, Matzen SH (2014) *Rectus abdominis diastasis*. J Plast Surg Hand Surg 48(3):163–169. https://doi.org/10.3109/2000656X.2013.859145
20. Chiarello CM, Falzone LA, McCaslin KE (2005) The effects of an exercise program on *diastasis recti abdominis in* pregnant women. J Womens Health Phys Ther 29(1):11–16
21. Mota PGFD, Pascoal AGBA, Carita AIAD et al (2015) Prevalence and risk factors of diastasis recti abdominis from late pregnancy to 6 months postpartum, and relationship with lumbo-pelvic pain. Manual Ther 20(1):200–205. https://doi.org/10.1016/j.math.2014.09.002
22. Liaw L-J, Hsu M-J, Liao C-F et al (2011) The relationships between inter-recti distance measured by ultrasound imaging and abdominal muscle function in postpartum women: a 6-month follow-up study. J Orthop Sports Phys Ther 41(6):435–443

23. Stokes IAF, Gardner-Morse MG, Henry SM (2010) Intra-abdominal pressure and abdominal wall muscular function: spinal unloading mechanism. Clin Biomech (Bristol, Avon) 25 (9):859–866. https://doi.org/10.1016/j.clinbiomech.2010.06.018
24. Pascoal AG, Dionisio S, Cordeiro F et al (2014) *Inter-rectus* distance in *postpartum* women can be reduced by isometric contraction of the abdominal muscles: a preliminary case-control study. Physiotherapy 100(4):344–348. https://doi.org/10.1016/j.physio.2013.11.006
25. Mannion AF, Pulkovski N, Toma V et al (2008) Abdominal muscle size and symmetry at rest and during abdominal hollowing exercises in healthy control subjects. J Anat 213(2):173–182
26. Therapeutic exercise for Lumbopelvic stabilization: a motor control approach for the treatment and prevention of low back pain. Churchill Livingstone, 2004, 1 p
27. Therapeutic exercise for spinal segmental stabilization in low back pain: scientific basis and clinical approach [Internet]. Churchill Livingstone, 1999. http://espace.library.uq.edu.au/view/UQ:145870
28. Benjamin DR, van de Water ATM, Peiris CL (2014) Effects of exercise on diastasis of the *rectus abdominis* muscle in the antenatal and postnatal periods: a systematic review. Physiotherapy 100(1):1–8. https://doi.org/10.1016/j.physio.2013.08.005
29. Mesquita LA, Machado AV, Andrade AV (1999) Physiotherapy for reduction of diastasis of the *recti abdominis* muscles in the *postpartum* period. Revista Brasileira de Ginecologia e Obstetrícia 21(5):267–272
30. Sheppard S (1996) The role of *transversus abdominus* in *postpartum* correction of gross divarication *recti*. Manual Ther 1(4):214–216
31. Keeler J, Albrecht M, Eberhardt L et al (2012) *Diastasis recti abdominis*: a survey of women's health specialists for current physical therapy clinical practice for postpartum women. J Womens Health Phys Ther 36(3):131–142
32. Mota P, Pascoal AG, Carita AI, Bø K (2015) The immediate effects on *inter-rectus* distance of abdominal crunch and drawing-in exercises during pregnancy and the *postpartum* period. J Orthop Sports Phys Ther 45(10):781–788
33. Noble E (1995) Essential exercises for the childbearing year: a guide to health and comfort before and after your baby is born. New Life Images, 1 p
34. Mantle J, Haslam J, Barton S (2004) Physiotherapy in obstetrics and gynaecology. Elsevier Health Sciences, 1 p
35. Boxer S, Jones S (1997) Intra-rater reliability of *rectus abdominis diastasis* measurement using dial calipers. Aust J Physiother 43(2):109–114
36. Hsia M, Jones S (2000) Natural resolution of *rectus abdominis diastasis*. Two single case studies. Aust J Physiother 46(4):301–307
37. Bø K, Berghmans B, Van Kampen M et al (2015) Evidence-based physical therapy for the pelvic floor: bridging science and clinical practice. ISBN: 978-0-7020-4443-4
38. Rankin G, Stokes M, Newham DJ (2006) Abdominal muscle size and symmetry in normal subjects. Muscle Nerve 34(3):320–326. https://doi.org/10.1002/mus.20589
39. Hodges PW, Pengel LHM, Herbert RD et al (2003) Measurement of muscle contraction with ultrasound imaging. Muscle Nerve W 27(6):682–692. https://doi.org/10.1002/mus.10375
40. Mendes Dde A, Nahas FX, Veiga DF et al (2007) Ultrasonography for measuring *rectus abdominis* muscles *diastasis*. Acta Cir Bras 22(3):182–186
41. Hides JA, Miokovic T, Belavý DL et al (2007) Ultrasound imaging assessment of abdominal muscle function during drawing-in of the abdominal wall: an intrarater reliability study. J Orthop Sports Phys Ther 37(8):480–486
42. Whittaker JL, Warner MB, Stokes MJ (2009) Induced transducer orientation during ultrasound imaging: effects on abdominal muscle thickness and bladder position. Ultrasound Med Biol 35(11):1803–1811. https://doi.org/10.1016/j.ultrasmedbio.2009.05.018
43. Whittaker JL, Warner MB, Stokes MJ (2010) Ultrasound imaging transducer motion during clinical maneuvers: respiration, active straight leg raise test and abdominal drawing in. Ultrasound Med Biol 36(8):1288–1297. https://doi.org/10.1016/j.ultrasmedbio.2010.04.020

44. Klimstra M, Dowling J, Durkin JL et al (2007) The effect of ultrasound probe orientation on muscle architecture measurement. J Electromyogr Kinesiol 17(4):504–514. https://doi.org/10.1016/j.jelekin.2006.04.011
45. Kwah LK, Pinto RZ, Diong J et al (2013) Reliability and validity of ultrasound measurements of muscle fascicle length and pennation in humans: a systematic review. J Appl Physiol 114 (6):761–769. https://doi.org/10.1152/japplphysiol.01430.2011
46. Bénard MR, Becher JG, Harlaar J et al (2009) Anatomical information is needed in ultrasound imaging of muscle to avoid potentially substantial errors in measurement of muscle geometry. Muscle Nerve 39(5):652–665. https://doi.org/10.1002/mus.21287
47. Noorkoiv M, Stavnsbo A, Aagaard P et al (2010) *In vivo* assessment of muscle fascicle length by extended field-of-view ultrasonography. J Appl Physiol 109(6):1974–1979. https://doi.org/10.1152/japplphysiol.00657.2010
48. Aggeloussis N, Giannakou E, Albracht K et al (2010) Reproducibility of fascicle length and pennation angle of *gastrocnemius medialis* in human gait *in vivo*. Gait Posture 31(1):73–77. https://doi.org/10.1016/j.gaitpost.2009.08.249
49. Mota P, Pascoal AG, Sancho F et al (2012) Test-retest and intrarater reliability of 2-dimensional ultrasound measurements of distance between *rectus abdominis* in women. J Orthop Sports Phys Ther 42(11):940–946. https://doi.org/10.2519/jospt.2012.4115

Biomechanical Analysis of the Damage in the Pelvic Floor Muscles During Childbirth

Dulce Oliveira, Marco Parente, Teresa Mascarenhas and Renato Natal Jorge

Abstract The vaginal birth is the leading cause of pelvic floor muscle injuries compromising its function, which can lead to pelvic organ prolapse, urinary incontinence, and other pelvic disorders. These conditions affect many women's quality of life. As such, biomechanical models emerge to analyze the impact of pregnancy and childbirth in the biomechanics of the pelvic floor, and determine features that potentially contribute to complications during vaginal delivery. Computer models allow structural hypotheses to be analyzed, such as the influence of the shape of the fetal head and its position at delivery, the consequence of specific obstetrical procedures, among others. Damage analysis is especially important to understand the pathophysiology of the associated dysfunctions. The continuous developments in imaging techniques, and the increased computing power, make possible for these frameworks to be clinically valuable, with customized computer models and subject-specific mechanical properties, both in useful time.

Keywords Computational model · Numerical analysis · Pelvic floor muscles
Pelvic floor dysfunction

D. Oliveira (✉) · M. Parente · R. N. Jorge
LAETA, INEGI, Faculty of Engineering, University of Porto, Porto, Portugal
e-mail: dulce.oliv85@gmail.com

T. Mascarenhas
Department of Gynecology and Obstetrics, Centro Hospitalar de São João—EPE, Faculty of Medicine, University of Porto, Porto, Portugal

S. Brandão et al. (eds.), *Women's Health and Biomechanics*,
Lecture Notes in Computational Vision and Biomechanics 29,
https://doi.org/10.1007/978-3-319-71574-2_11

Highlights

- Pregnancy and vaginal delivery are considered risk factors for developing pelvic floor dysfunction
- Strategies to reduce or prevent such condition are focused on course of pregnancy, mode of delivery, management of the delivery, as well as in pelvic physiotherapy
- Computational modeling is used to analyze the impact of vaginal delivery in the pelvic floor muscles by quantifying physical/mechanical variables (stress, strain, forces, damage, pressure)
- Childbirth simulations are performed considering different scenarios and analyzing several structural hypotheses, observing how they affect the outcomes
- Biomechanical models can be used to prevent birth-related debilitating conditions, to optimize the design of implanted medical devices used to correct them, and to develop decision support systems that can provide guidance for healthcare professionals.

1 Introduction

The pelvic floor is a physiologically intricate and structurally complex region composed of the pelvic floor muscles (PFM), connective tissues, and the pelvic organs. The PFM provides structural support to the organs of the pelvic cavity by keeping them in their anatomical position, and also collaborate in the continence mechanism by closing the pelvic openings. Damage to these anatomical relationships may lead to pelvic floor dysfunction (PFD). The main symptoms include urinary or fecal incontinence, pelvic organ prolapse, sexual dysfunction or pelvic pain. PFD represents an important health problem, affecting approximately 50% of women older than 50 years [1]. The major risk factor for developing these conditions is vaginal birth due to pelvic floor injuries for which physiological changes (hormonal changes and other stressors) also contribute [2]. During vaginal delivery, the PFM can stretch beyond a permissible length, damaging them both mechanically and its nervous control [3], enhanced by prolonged second stage of labor and fetal macrosomia [4, 5]. These PFM defects are easily detected and evaluated by ultrasound [6] and magnetic resonance imaging (MRI) [7, 8]. However, further research is needed to understand if this type of trauma is common or severe enough to require a change in clinical practice.

Biomechanics is considered one of the main topics of current pelvic floor research. Increasingly detailed and sophisticated biomechanical modeling is a useful approach for understanding PFM function, and to explore vaginal birth and PFD mechanisms. These models are vital to estimate the mechanical changes on PFM during pregnancy and vaginal delivery, assess failures and evaluate sequelae of vaginal birth, especially because in vivo experimental work cannot be performed due to clinical, technical and ethical reasons. The translation of biomechanics research to the clinical settings may contribute to better understand the etiology of these complex conditions, and the mechanisms of birth-related injuries, improve assessment and treatment of PFD, assume preventive strategies and optimize surgical procedures.

2 The Mechanics of Childbirth

Labor is a physiologic process during which the products of conception, namely the fetus, amniotic fluid, placenta, and membranes are expelled from the uterus. The forces for fetal expulsion are from the uterine contractions (triple descending gradient) and the abdominal contractions (intrauterine pressure increase). During the cardinal movements, the fetus successfully negotiates the pelvis, involving changes in the position of the fetus's head during labor, depending on the driving force, the birth canal and the fetus size. Although labor and delivery occur continuously, the cardinal movements are described in the following sequence (in relation to a cephalic presentation): engagement, descent, flexion, internal rotation, extension, restitution and external rotation, and expulsion. The labor is a process divided into three stages. The first stage of labor begins with regular uterine contractions and ends with complete cervical dilation. The second stage ranges from full dilation of the cervix to the delivery of the fetus. Finally, the third stage of labor covers the period between the delivery of the fetus and the delivery of the placenta and fetal membranes [9].

Regarding computational simulations, research groups mainly focus on the second stage of labor, as is the phase in which tissues are most demanded in order to allow the descent of the fetal head, suffering large deformations. Potential injuries to the PFM can occur during this process depending on their morphology and mechanical properties, morphology and adaptability of the fetal head, type of delivery and medical intervention, among others [10–13]. It is crucial at this stage to assess the relationship between the fetal head and the maternal pelvic structures (e.g., the shape of the pelvic bones) to predict birth outcomes and outline the strategy to be adopted in case of difficulties (episiotomy, instrumented delivery or cesarean section).

3 Impact of Vaginal Delivery on Biomechanics of Pelvic Floor

During pregnancy and labor, the pelvic floor is under remarkable biomechanical changes, partly due to hormone-mediated physiological changes. Animal models have shown that the vaginal wall and supporting tissues are more distensible and less stiff during pregnancy, being weaker and less tolerant to stress [14]. These changes, in turn, renders tissues more susceptible to trauma. Therefore, when the fetus is progressively descending through the birth canal, the head induces substantial stretch/compression in the pelvic floor structures—especially the PFM—easily exceeding their load bearing, which may lead to tissue damage.

These injuries in the PFM can be identified by MRI and three-dimensional ultrasound. During vaginal birth, the *pubococcygeus* muscle undergoes a stretch ratio of 3.26, according to a biomechanical simulation conducted by Lien et al. [15]. In addition, the substantial stresses and strains occurring during childbirth increase collagenase activity [16] and impacted the turnover of vaginal elastin, which are critical components of vaginal and pelvic floor connective tissues [14]. These events may confer inferior mechanical properties to the tissues, increasing distensibility and decreasing stiffness and maximal stress.

According to literature, 20 to 36% of women experience significant PFM injury at the time of vaginal birth [7, 17], with threefold odds for requiring instrumented vaginal delivery [13, 18]. The damage can generate severe and sometimes permanent degradation of the mechanical properties of pelvic organs and PFM, leading to PFD [14, 19]. Different symptoms may occur depending on the portion of the PFM affected, due to loss of a specific mechanical function. Therefore, understanding the biomechanics of the PFM and determining the biomechanical impact of parturition on the pelvic floor is essential to disclose the pathophysiology of PFD induced by vaginal delivery [20]. Furthermore, longitudinal studies are needed to clarify the role of these injuries and PFM function in the later development of PFD.

4 Biomechanical Computational Models for Simulation of Vaginal Delivery

The biomechanical aspects of childbirth are still not completely understood; as such, computational models come with the intention of enhancing the current knowledge by studying the mechanical aspects during vaginal delivery, such as stress, strain, forces, and contact pressures. As reported before, clinical, technical and ethical reasons hinder in vivo experimental tests, resorting to biomechanical models estimating the mechanical changes on the PFM during delivery [19, 21], quantifying muscle damage [12], analyzing the fetus descent [22], the effect of the fetal head molding on the PFM mechanical behavior [23], and also the influence of several obstetric procedures, such as episiotomy (Fig. 1) [24].

In the following sections, the main steps necessary to develop numerical models capable of simulating vaginal delivery are presented.

4.1 Geometrical Models of the Pelvic Floor Structures

To accurately model the behavior of the PFM during the second stage of labor, a precise anatomical representation of the pelvic floor is necessary. The creation of detailed anatomical models is mainly based on high-resolution medical imaging, and most biomechanical models are based on MRI scans from nulliparous women [22, 25–27]. Quantitative descriptions from cadaver-based studies were also applied

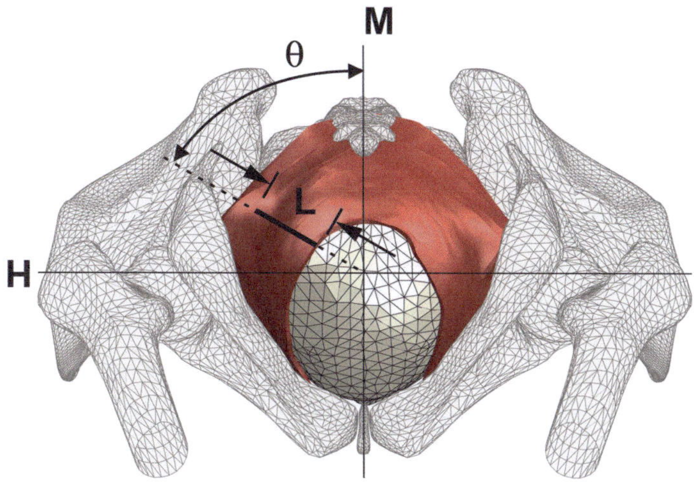

Fig. 1 Inferior view of the female pelvic outlet, during a vaginal delivery, with the pelvic bones, partial femurs and the PFM (red colored) to schematize a mediolateral episiotomy incision, identifying the episiotomy angle, θ, and the incision length, L. The incision begins in the midline, M, in the inferior portion of the hymeneal ring (intersection of the midline and the horizontal line, H). The childbirth simulations performed should optimize the influence of episiotomy finding the optimal value for the variables

[21]. The most numerical models consider the PFM as a single structure, identifying only the various portions of the muscle [19, 21, 26–28]. However, there are also studies considering individual muscle bands [15]. Furthermore, lateral and posterior insertions of the PFM to the bony pelvis were provided through deformable supportive structures continuous with the PFM [21, 28, 29] or linear springs [19, 26]. Some models have included also the coccyx (to allow posterior deformation) [26, 27], symphysis pubis [28, 29], perineal body (to constrain the caudal opening of the PFM) [26], and bony pelvis [19, 21]. Pelvic bones usually do not participate in the simulation, being considered rigid and immovable, but are important in helping to define the boundaries of the pelvic floor.

In addition to the pelvic structures anatomy, imaging techniques have also been used to visualize muscle fiber architecture through diffusion MRI [30]. However, the application of this technique to in vivo imaging of the PFM fibers has not yet been validated, remaining hypothetical.

Childbirth models still do not include other pelvic organs such as the bladder, rectum, and vagina. As such, its contribution to the labor mechanics remains unknown.

4.2 Geometrical Models of the Fetus

Medical imaging of the fetal head at later stages of gestation should be used to create the fetal skull model, however, due to ethical restrictions, such information is scarce. As such, fetal head models have been approximated to spheres [15, 19], or created based on CT scans of newborn/stillborn heads [21, 23].

Regarding fetal head movements during its descent through the birth canal, some researchers have prescribed the path of descent [21], while others have allowed that this movement arises from the interactions between the fetal head, bony pelvis, and the PFM [23]. This approach allows that minimal mechanical energy state is established at every step of the descent.

In order to more accurately estimate the forces and muscle deformation during labor, the fetal head molding should be accounted for. However, this increases the complexity of the models, as well as originates convergence problems and, as such, has been neglected in most modeling frameworks.

4.3 Constitutive Modeling of the Pelvic Floor Muscles

To create a precise and realistic numerical model of childbirth biomechanics, accurate constitutive models are necessary to characterize the mechanical behavior of the PFM tissue. Depending on the application, a material model can vary in complexity. The constitutive models used to describe the behavior of the PFM vary among research groups, being assumed as isotropic, transversely isotropic or anisotropic, and elastic, hyperelastic or visco-hyperelastic [19, 21, 26, 31]. While some authors account for the effect of the variable time [26], others still account for the damage effect [12].

In addition to the constitutive law that enables estimating the material parameters of such constitutive relations, quantitative experimental data on the material properties of the PFM tissue during delivery should be performed. However, due to ethical constraints, extrapolation is typically required. Experimental data at this stage would be important since the mechanical properties of the tissues are affected by the hormonal *status* during pregnancy and childbirth [31], as observed in animal models [14]. To overcome these limitations, new technologies should be tested such as inverse finite element analysis [32]. Although elastography-based imaging techniques for non-invasive evaluation of tissue mechanical properties are in vogue, their use for analyzing PFM is limited since they assume the tissue under analysis as linear, elastic, isotropic and incompressible [33].

5 Childbirth-Related Pelvic Floor Trauma

Birth-related pelvic floor injury can be caused by muscle damage from compression or tearing, and neuropathy from compression or stretch, due to acute physical strains during childbirth and intraperitoneal forces. Such trauma affects mainly the puborectalis muscle, although in some women, may also involve the iliococcygeus muscle. The amount of time that the muscle structure is undergoing compression can also lead to muscle trauma, hence a second stage duration equal to or greater than 110 min is a known risk factor for pelvic floor injury. Another risk factor is the size of the head circumference (≥ 35.5 cm) [5]. Computational simulations of childbirth can contribute to better understand the origin of the injuries, allowing preventive actions timelier and effectively. In addition, these computer models can help to set the maximum force that would not induce damage and also, to relate the fetal head circumference to the geometry of the pelvis to anticipate trauma during vaginal delivery.

Modeling the PFM with a material model that describes the mechanical behavior in the failure region allows studying the damage mechanics, contributing to determine the conditions for the start of the first damage event, predicting its evolution, and characterizing, quantifying and analyzing its effect on the mechanical response of the material [12]. The results can then be compared with actual cases of PFM trauma identified with pelvic floor imaging.

Determining the impact of childbirth on the PFM biomechanics is important to improve the knowledge regarding the pathogenesis of PFM injury and to identify factors that distinguish women who suffer birth-related PFM trauma from those who do not. Childbirth-related pelvic floor damage occurs and commonly accompanies even a seemingly uneventful childbirth, as verified through computational simulations (Fig. 2) [12], being also visible through imaging [7]. However, it remains to be realized whether this is an event that leads to PFD. These biomechanical models can contribute to predict the long-term consequences of PFM injury, since they remain uncertain.

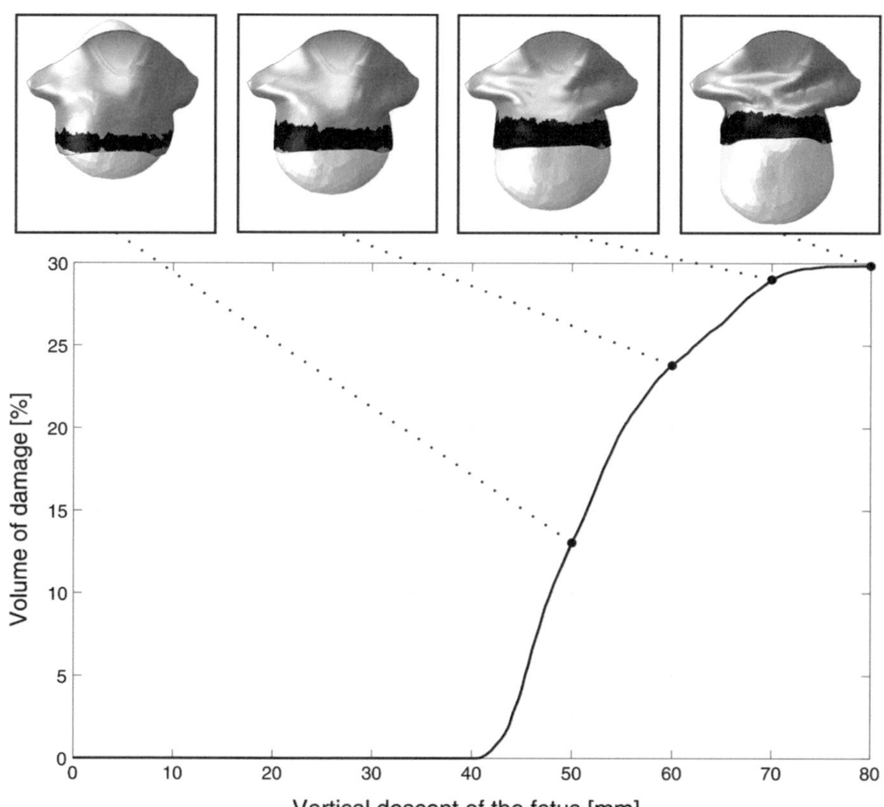

Fig. 2 Damage evolution based on fetus descent and fiber damage contour plot (damaged elements black colored) corresponding to fetus vertical displacements of 50, 60, 70, and 80 mm

6 Conclusions

The biomechanical models developed so far are very promising in the representation of the anatomy and mechanical behavior of the PFM during vaginal delivery, and may signify an important noninvasive tool to prevent and study the origin of PFD and to improve the preparation of surgeries. Vaginal delivery simulations intend to help understand the mechanisms that trigger the dysfunctions and assist in the identification of risk groups.

Modeling childbirth is a highly complex issue, from the modeling of the complex geometries involved (mother's pelvis and fetus) up to the characterization of the mechanical behavior of each tissue engaged, going through the clinical validation. Most research teams are especially interested in the PFM, therefore, computational models of childbirth, typically, do not include other pelvic organs. Mainly because the PFM is the structure more affected by the vaginal delivery and the inclusion of various structures would further increase the complexity of the simulation, and also because the remaining pelvic structures are highly compliant, offering little resistance to the passage of the fetus. The constant developments in imaging techniques, computational analysis methods, along with increased computing power, and the development of noninvasive biomechanical measurement methods, offer the promise of major developments in the field of childbirth biomechanics. In the future, subject-specific geometrical models can be created based on imaging of pre-pregnant women and adjusted to each case of pregnancy, determining the optimal delivery strategy that minimizes the risk of childbirth-related pelvic floor injuries. By simulating in a realistic way a vaginal delivery, these models will become valuable tools for obstetricians.

References

1. Wilson L, Brown JS, Shind GP et al (2001) Annual direct cost of urinary incontinence. Obstet Gynecol 98(3):398–406. https://doi.org/10.1016/S0029-7844(01)01464-8
2. Goldberg RP (2007) Effects of pregnancy and childbirth on the pelvic floor. In: Culligan PJ, Goldberg RP (eds) Urogynecology in primary care. Springer, London, pp 21–33
3. Bortolini MAT, Drutz HP, Lovatsis D et al (2010) Vaginal delivery and pelvic floor dysfunction: current evidence and implications for future research. Int Urogynecol J 21 (8):1025–1030. https://doi.org/10.1007/s00192-010-1146-9
4. Baracho SM, Barbosa da Silva L, Baracho E et al (2012) Pelvic floor muscle strength predicts stress urinary incontinence in primiparous women after vaginal delivery. Int Urogynecol J 23:899–906. https://doi.org/10.1007/s00192-012-1681-7
5. Valsky DV, Lipschuetz M, Bord A et al (2009) Fetal head circumference and length of second stage of labor are risk factors for *levator ani* muscle injury, diagnosed by 3-dimensional transperineal ultrasound in primiparous women. Am J Obstet Gynecol 201(1):91.e1–91.e7. https://doi.org/10.1016/j.ajog.2009.03.028
6. Liu F, Xu L, Ying T et al (2014) Three-dimensional ultrasound appearance of pelvic floor in nulliparous women and *postpartum* women one week after their first delivery. Int J Med Sci 11(3):234–239. https://doi.org/10.7150/ijms.7384

7. DeLancey JOL, Kearney R, Chou Q et al (2003) The appearance of *levator ani* muscle abnormalities in magnetic resonance images after vaginal delivery. Obstet Gynecol 101 (1):46–53. https://doi.org/10.1016/S0029-7844(02)02465-1
8. van Delft K, Thakar R, Sultan A et al (2014) *Levator ani* muscle avulsion during childbirth: a risk prediction model. BJOG 121(9):1155–1163. https://doi.org/10.1111/1471-0528.12676
9. FIGO Safe Motherhood and Newborn Health (SMNH) Committee (2012) Management of the second stage of labor. Int J Gynaecol Obstet 119:111–116. https://doi.org/10.1016/j.ijgo.2012.08.002
10. Braekken IH, Majida M, Engh ME et al (2014) Are pelvic floor muscle thickness and size of *levator hiatus* associated with pelvic floor muscle strength, endurance and vaginal resting pressure in women with pelvic organ prolapse stages I-III? A cross sectional 3D ultrasound study. Neurourol Urodyn 33(1):115–120. https://doi.org/10.1002/nau.22384
11. Handa VL, Blomquist JL, McDermott KC et al (2012) Pelvic floor disorders after vaginal birth: effect of episiotomy, perineal laceration, and operative birth. Obstet Gynecol 119(2 Pt1):233–239. https://doi.org/10.1097/AOG.0b013e318240df4f
12. Oliveira DA, Parente MPL, Calvo B et al (2016) Numerical simulation of the damage evolution in the pelvic floor muscles during childbirth. J Biomech 49(4):594–601. https://doi.org/10.1016/j.jbiomech.2016.01.014
13. Albrich S, Laterza R, Skala C et al (2012) Impact of mode of delivery on levator morphology: a prospective observational study with three-dimensional ultrasound early in the postpartum period. BJOG 119(1):51–61. https://doi.org/10.1111/j.1471-0528.2011.03152.x
14. Rahn DD, Ruff MD, Brown SA et al (2008) Biomechanical properties of the vaginal wall: effect of pregnancy, elastic fiber deficiency, and pelvic organ prolapse. Am J Obstet Gynecol 198(5):590.e1–590.e6. https://doi.org/10.1016/j.ajog.2008.02.022
15. Lien K-C, Mooney B, DeLancey JOL et al (2004) *Levator ani* muscle stretch induced by simulated vaginal birth. Obstet Gynecol 103(1):31–40. https://doi.org/10.1097/01.AOG.0000109207.22354.65
16. Zong W, Jallah ZC, Stein SE et al (2010) Repetitive mechanical stretch increases extracellular collagenase activity in vaginal fibroblasts. Female Pelvic Med Reconstr Surg 16(5):257–262. https://doi.org/10.1097/SPV.0b013e3181ed30d2
17. Dietz HP, Lanzarone V (2005) Levator trauma after vaginal delivery. Obstet Gynecol 106 (4):707–712. https://doi.org/10.1097/01.AOG.0000178779.62181.01
18. Chan SSC, Cheung RYK, Yiu AK et al (2012) Prevalence of *levator ani* muscle injury in Chinese women after first delivery. Ultrasound Obstet Gynecol 39(6):704–709. https://doi.org/10.1002/uog.10132
19. Hoyte L, Damaser MS, Warfield SK et al (2008) Quantity and distribution of *levator ani* stretch during simulated vaginal childbirth. Am J Obstet Gynecol 199(2):198.e1–198.e5. https://doi.org/10.1016/j.ajog.2008.04.027
20. Oliphant SS, Nygaard IE, Zong W et al (2014) Maternal adaptations in preparation for parturition predict uncomplicated spontaneous delivery outcome. Am J Obstet Gynecol 211 (6):630.e1–630.e7. https://doi.org/10.1016/j.ajog.2014.06.021
21. Parente MPL, Natal Jorge RM, Mascarenhas T et al (2008) Deformation of the pelvic floor muscles during a vaginal delivery. Int Urogynecol J Pelvic Floor Dysfunct 19(1):65–71. https://doi.org/10.1007/s00192-007-0388-7
22. Buttin R, Zara F, Shariat B et al (2013) Biomechanical simulation of the fetal descent without imposed theoretical trajectory. Comput Methods Programs Biomed 111(2):389–401. https://doi.org/10.1016/j.cmpb.2013.04.005
23. Silva MET, Oliveira DA, Roza TH et al (2015) Study on the influence of the fetus head molding on the biomechanical behavior of the pelvic floor muscles, during vaginal delivery. J Biomech 48(9):1600–1605. https://doi.org/10.1016/j.jbiomech.2015.02.032
24. Oliveira DA, Parente MPL, Calvo B et al (2016) A biomechanical analysis on the impact of episiotomy during childbirth. Biomech Model Mechanobiol 15(6):1523–1534. https://doi.org/10.1007/s10237-016-0781-6

25. Li X, Kruger JA, Nash MP et al (2011) Anisotropic effects of the levator ani muscle during childbirth. Biomech Model Mechanobiol 10(4):485–494. https://doi.org/10.1007/s10237-010-0249-z
26. Jing D, Ashton-Miller JA, DeLancey JOL (2012) A subject-specific anisotropic visco-hyperelastic finite element model of female pelvic floor stress and strain during the second stage of labor. J Biomech 45(3):455–460. https://doi.org/10.1016/j.jbiomech.2011.12.002
27. Berardi M, Martinez-Romero O, Elias-Zuniga A et al (2014) *Levator ani* deformation during the second stage of labour. Proc Inst Mech Eng H 228(5):501–508. https://doi.org/10.1177/0954411914533678
28. Yan X, Kruger JA, Nielsen PMF et al (2015) Effects of fetal head shape variation on the second stage of labour. J Biomech 48(9):1–7. https://doi.org/10.1016/j.jbiomech.2015.02.062
29. Li X, Kruger JA, Nash MP et al (2010) Effects of fetal head motion on pelvic floor mechanics. Computational biomechanics for medicine. Springer, New York, pp 129–137
30. Zijta FM, Froeling M, van der Paardt MP et al (2011) Feasibility of diffusion tensor imaging (DTI) with fibre tractography of the normal female pelvic floor. Eur Radiol 21(6):1243–1249. https://doi.org/10.1007/s00330-010-2044-8
31. Li X, Kruger JA, Nash MP et al (2010) Effects of nonlinear muscle elasticity on pelvic floor mechanics during vaginal childbirth. J Biomech Eng 132(11):111010. https://doi.org/10.1115/1.4002558
32. Silva MET, Brandão S, Parente MPL et al (2016) Establishing the biomechanical properties of the pelvic soft tissues through an inverse finite element analysis using magnetic resonance imaging. Proc Inst Mech Eng H 230(4):298–309. https://doi.org/10.1177/0954411916630571
33. Sigrist R, Liau J, Kaffas A et al (2017) Ultrasound elastography: review of techniques and clinical applications. Theranostics 7(5):1303–1329. https://doi.org/10.7150/thno.18650

Part IV
Clinical Approach on the Female Pelvic Floor

Pelvic Floor in Female Athletes: From Function to Dysfunction

Alice Carvalhais, Thuane Da Roza and Cinara Sacomori

Abstract Despite the benefits derived from exercise practice, some adverse effects have been described in female athletes. Regarding the problems associated with the pelvic floor, the most commonly reported is urinary incontinence (UI). However, menstrual irregularity, anal incontinence, sexual dysfunction, and eating disorders are also demonstrated among athletes. Practicing sports always involves increased intra-abdominal pressure, but the wide prevalence range of UI found between each sport suggests that some sports could be associated with higher risk than others. Regarding the relation between exercise and pelvic floor dysfunctions, two opposite theories have been suggested: (i) female athletes have strong pelvic floor muscles (PFM); and (ii) female athletes may overload, stretch, and weaken the PFM. Both theories are supported by the results of different studies, suggesting that the mechanism behind exercise-induced UI is not yet completely understood. Rather than a single factor, it is more likely that several environmental and individual factors interact to contribute to pelvic floor dysfunctions. UI negatively affects the quality of life of incontinent women in several domains, which reinforces the need to implement preventive strategies in order to counteract the negative effects of exercise.

Keywords Athletes · Pelvic floor muscles · Urinary incontinence
Quality of life

A. Carvalhais (✉)
Department of Physiotherapy, CESPU—North Polytechnic Institute of Health, Paredes, Portugal
e-mail: alice.carvalhais@ipsn.cespu.pt

T. Da Roza
Department of Physiotherapy, Santa Catarina State University, Florianópolis, Brazil

C. Sacomori
School of Kinesiology, Universidad Bernardo O'Higgins, Santiago de Chile, Chile

S. Brandão et al. (eds.), *Women's Health and Biomechanics*,
Lecture Notes in Computational Vision and Biomechanics 29,
https://doi.org/10.1007/978-3-319-71574-2_12

Highlights

- Excessive training increases the risk of specific injuries in female athletes
- Pelvic floor muscles are affected by high impact sports, which leads to the development of several dysfunctions, most commonly urinary incontinence
- Some modifiable risk factors for urinary incontinence have been identified, which can help to develop more effective preventive interventions
- Since the mechanism behind exercise-induced incontinence is not completely clear, there is a need to perform more studies to clarify the pathophysiology of pelvic floor dysfunction due to the practice of physical exercise.

1 Introduction

Usually, an athlete is regarded as a healthy and fit person. However, the terms "fit" and "healthy" have entirely separate meanings and have separate definitions: fitness describes the ability to perform a specific physical task, which includes exercise and sports; whereas health means a state of complete mental, social, and physical well-being, where all bodily systems (nervous, hormonal, immune, digestive, etc.) function in harmony [1]. Most elite athletes push themselves beyond the "healthy" limit [2], for instance, practicing and competing with pain or not allowing a complete recovery from injuries and illnesses [3], which can lead to overtraining or overreaching conditions.

It is well known that endocrine [4] and psychological disruptions [4], or physical changes [5], may occur due to excessive training. Moreover, sports practice can increase the risk of some specific injuries in female athletes. Among these risks, in the last decade, studies have reported that intense exercise may lead to gynecological problems, since exercise can affect the normal menstrual cycle in several forms (primary and secondary amenorrhea, oligomenorrhea, short luteal phases, and anovulation). A possible explanation for this is the decrease in gonadotrophin-releasing hormone pulses from the hypothalamus, which consequently will shut down stimulation of the ovaries. In addition, athletes with low weight have the risk of developing the female athlete triad: amenorrhea, osteoporosis, and eating disorder [6]. Poor nutrition in conjunction with inappropriate volume of training can put athletes at higher risk to develop musculoskeletal injuries [1]. Beyond that, the increase in intra-abdominal pressure generates a downward force in the pelvic floor muscles (PFM). Bø and Stein showed that there is coactivation of the abdominal muscles and PFM in functional activities such as head and shoulder raising [7], which afterward was supported by Sapsford and colleagues' findings [8]. They all highlight the functional interdependence between the different structures that limit the abdominal cavity.

The abdomen wall is that area below the diaphragm, the upper portion referred to as the abdomen proper and the lower portion being the lesser pelvis [9]. Therefore, the relationship between abdominal muscles and PFM is much debated. Studies have demonstrated that if the musculature of the pelvic floor is weak in relation to surrounded musculature of the abdominal cavity, it cannot co-contract effectively and urinary loss episodes can occur, especially when women are submitted to physical exertion, even in the absence of other risk factors.

Regarding sports activities, the ground reactions forces transmitted to the PFM during running (11 km/h) can reach 2.5 times the body weight [10] and it seems that PFM activity increases with speed, ranging from 46.2 to 92.0% maximal voluntary contraction at 7 km/h, from 55.7 to 107.5% at 11 km/h and 80.8–151.8% at 15 km/h [11]. Landing (also called high impact) may be an important factor in effecting functional and morphological changes in the PFM [12] due to the frequent, sudden, and intense increases in intra-abdominal pressure. The voluntary and involuntary PFM contractions associated with the central and peripheral control mechanisms are the main causes for the maintenance and proper function of the continence mechanism. If these systems do not work in harmony, pelvic floor dysfunctions may appear. Urinary incontinence (UI), anal incontinence, dyspareunia, and pelvic organ prolapse are the most common disorders [13].

In this chapter, we will describe the main pelvic floor dysfunctions in female athletes, and explain why they are affected by this problem and the impact it has on their quality of life.

2 Pelvic Floor (Dys)Function in Athletes

According to the International Urogynecological Association (IUGA) and to the International Continence Society (ICS) Joint Report, UI is the complaint of involuntary loss of urine [14], and this is the most common pelvic floor dysfunction among athletes. Its prevalence ranges from 0% [15] to 80% [16], with the highest prevalence among high-impact sports practitioners.

In 1994, Nygaard et al. studied 144 female athletes, from nine different sports, and found that golf was the one presenting 0% of prevalence of UI, while gymnastics and tennis (67% and 50%, respectively) were the sports with the highest prevalence [15]. Considering that the PFM are exposed to high forces during landing, Eliasson et al. (2002) studied the prevalence of UI in elite trampolinists, finding that 80% of those athletes reported UI during trampoline training [16]. Recently, these results were further confirmed by Da Roza et al. (2015) [17] , which found a prevalence of 72.7% of UI in trampolinists.

Athletes from others sports also present a wide range of prevalence for UI, as for example, 43% for ballet, aerobics 40%, badminton 31%, volleyball 30%, athletics 25%, handball 21%, and basketball 17% [18]. Another study compared three sports —athletics, basketball, and indoor football—to verify if the prevalence of UI is similar among them. The results showed that the prevalence across the 3 sports was similar and was not affected by age [19]. In addition, Bø and Borgen found that the prevalence of UI within sport groups varied between 37.5% (weight and technical sports) and 52% (aesthetic sports), however without a significant difference between them [20]. Despite the fact that sports practice itself involves increased intra-abdominal pressure, the wide range of UI found between the sports suggests that some sports could be riskier than others.

Currently, other pelvic floor dysfunctions are reported among athletes. A study assessed the prevalence of lower urinary tract symptoms and found that 13.3% present dysuria, and that urinary straining was present in 27.8% of the athletes [21]. Regarding symptoms of anal incontinence, sexual dysfunctions, and prolapse, a recent study compared athletes with nonathletes. The authors found that an involuntary loss of gas was the only symptom associated with the posterior pelvic compartment reported by the participants (athletes: 64.6% (n = 96) and nonathletes: 58.5% (n = 67), p = 0.438). The occurrence of dyspareunia was reported by 13.8% of athletes and by 21.9% of nonathletes with p = 0.351. Concerning vaginal laxity, 13.8% of the athletes, and 19.2% of the nonathletes complained of this symptom (p = 0.520), while pelvic organ prolapse was only reported by nonathletes [13]. In the same line, Vitton et al. (2011) compared two groups of young women: one practicing intensive (>8 h weekly), versus non-intensive sport practice. The authors found that the most common symptom of anal incontinence was flatus (63.6% non-intensive sport practice; 84% intensive sport practice) and after adjustment, practice intensive sport was a significant risk factor for anal incontinence (OR 2.99, 95% CI 1.29–6.87, p = 0.010) [22]. Moreover, prevalence of UI and dyspareunia were higher and significantly different in the sport practice group.

3 Why Does Pelvic Floor Dysfunction Occur in Athletes

Before we explain the pelvic floor reaction to exercises, it is important to clarify what intra-abdominal pressure is, and how it can be altered. Intra-abdominal pressure is the pressure concealed within the abdominal cavity [23] that oscillates according to the interaction between the diaphragm, abdominal muscles, low back spine, and PFM [8]. Thus, whenever a person exercises, these regions are solicited and tensed. This tension is usually made unconsciously with a change in respiratory pattern and in conjunction with a contraction of the abdominal wall, leading to an increase in intra-abdominal pressure. With the evolution of the human being, going from quadruped to erect posture, the PFM not only needs to counteract against the forces of intra-abdominal pressure but also the inertia and gravity forces.

There are two opposing theories about the pelvic floor in athletes: (i) female athletes have strong PFM; and (ii) female athletes overload, stretch and weaken their PFM [24]. Since the PFM and abdominal muscles have a co-contraction [8], the first hypothesis suggests that at each increase in the intra-abdominal pressure, the PFM will be trained, and therefore should be stronger and hypertrophied. In this context, a study compared PFM morphology using translabial three-dimensional and four-dimensional (3D/4D) ultrasonography between athletes and controls [12]. The findings showed that the athletes group has significantly larger diameter of pubovisceral muscle, large area of the *levator hiatus* during a voluntary Valsalva maneuver and have greater bladder neck descent when compared to controls [12].

In the same line, Da Roza et al. (2015), while studying a group of athletes, demonstrated that despite the fact that incontinent athletes evidenced thicker pubovisceral muscles (at the level of the midvagina) than the continent ones, no difference in PFM strength assessed by Oxford Scale was found [25]. A possible explanation for these finding is that UI can occur due to a decreased response or delayed reaction of the pubovisceral muscle rather than just being the result of its morphology (thickness) and strength [25].

The opposite theory is explained by different principles; biomechanics could be one of them, since the pelvis is located in the lower part of the trunk and connects the axial skeleton to the lower limbs. This region is responsible for transferring the trunk forces and the ground reaction loads between the legs and the vertebral column [9]. Hay (1993) describes that the maximum vertical ground reaction forces during running can reach 3 to 4 times the body weight and 5 to 12 times during jumping. Recently, research has reported that running speed can also affect the activation of the PFM [26]. According to Nygaard et al. (1996), how impact forces are absorbed may be a potential aetiology for UI. Accordingly, the authors investigated the relation between UI and force absorption on impact, as assessed by foot arch flexibility [27] and found a significant association between decreased foot flexibility (medial longitudinal arch height in two gait stances: neutral and maximally dorsiflexed ankle position) and UI among 47 elite athletes practicing five different sports [27]. The results suggest that how incontinence in athletes may be related to impact forces are transmitted to the pelvic floor.

Recent research using electromyography has reported that running speed can also affect the activation of the PFM [28]. Running at 11 km/h generates further activation in PFM when compared to 7 and 9 km/h [28]. In addition, running appears to trigger pre-activation before heel strike and reflex activation after heel strike [11], suggesting that PFM not only requires strength but also speed of contraction. This response to the rise of intra-abdominal pressure is due to distension of muscle spindles within PFM that further generates a co-contraction of the PFM. In this sense, it might be speculated that this demand to the PFM could have a negative impact on intrafusal fibers, diminishing their responsiveness to stretching, which may result in a delay to contraction [17].

PFM fatigue might be another reason to provoke UI in athletes. The fact that athletes leak during exercise and that they report urine loss more frequently during the second part of the training session and the second part of competition [29], associated with findings that training volume has significant associations with the severity of incontinence [17], may indicate that the mechanism of continence control is debilitated after repetitive efforts [30]. Furthermore, Ree and colleagues demonstrated a reduction of 20% in mean maximal voluntary contraction pressure after strenuous exercise, suggesting development of short-term muscle fatigue [31].

Recently, researchers have attempted to determine whether training in different weather conditions, such as temperature and humidity, might also influence the prevalence of UI [32]. To this end, the authors compared Polish female cross-country skiers and runners, but no statistically significant differences between both groups on the UI symptoms were found [32].

4 The Impact on Quality of Life

It is well known that UI negatively affects the quality of life (QOL) of incontinent women in several domains [33, 34], such as social and sexual relationships and self-perception, making them feel anxious and embarrassed [35]. Moreover, UI can be a barrier to exercise, leading practitioners to exercise less, change the type of exercise or even abandon the sport practice [36].

Regarding the overall impact of UI in athletes, few reported the condition as a moderate or severe problem [20]. In fact, it has been referred that athletes self-reported the condition as "not bothered" or interfering "mildly" [17, 32] and only a low percentage (21.1%) considered urine leakage as a hygienic problem [18]. Contrary to this, other authors found that athletes experienced feelings of concern, frustration [19], embarrassment, anxiety, and fear [15] due to UI. However, despite their concerns about the urine loss, the athletes stated that the condition had no current impact on their daily lives [19]. This fact may be explained because the amount of urine loss often is small [21] and in some cases, the majority of the athletes have losses only during sport practice [22].

Some athletes aim to hide the loss by wearing a pad [18, 22], while others try to reduce the volume of bladder urine during the exercise practice, either reducing the fluid intake [19] or emptying the bladder before training [13, 37]. Interestingly, the majority of the athletes with UI never discussed the occurrence of urine leakage with anyone or sought treatment [38]. Even among elite athletes, who are closely followed by a medical team, few report the loss of urine to their physician or other health professionals [15, 18]. Margalith et al. demonstrated that shame was the major cause for hiding their condition and to delayed seeking help among younger women, while fear of surgery was the cause for the older ones [39]. Despite the small amount of urine leakage and that they stated the condition as "not bothered" or interfering "mildly" in their lives, they assumed urinary leakage as affecting their performance [19].

It has been reported that a large number of young women do not have any knowledge either about pelvic floor function/dysfunction or UI prevention and treatment [40], which can be an obstacle to the early identification of the problem and to the application of adequate treatment that allows to solve the problem effectively. Given the increasing number of women practicing exercise and the growing evidence of association between UI and sport practice, there is a need to develop educational campaigns in order to prevent UI in females practicing sports.

5 Conclusions

We can conclude that being an athlete is not synonymous of being a healthy person. Athletes are included in a specific population, who sometimes train beyond their limits to achieve success. The highest prevalence rates of UI are registered in high impact sports, even in women at younger ages. Incontinent athletes state assume that this dysfunction negatively affects their quality of life in several domains, including their athletic performance; however, despite that, few look for medical advice.

There are several identified risk factors for urinary incontinence and some of them are modifiable. Thus, it is possible to prevent or minimize damages of this dysfunction. Currently, there is growing evidence that women enrolled in exercise practice are at major risk to develop pelvic floor dysfunction, and UI is the most common. Notwithstanding the increasing number of studies related to this condition, current understanding of the pathophysiology remains incomplete.

References

1. Maffetone PB, Laursen PB (2016) Athletes: fit but unhealthy? Sports Med Open 2(1):24
2. O'Keefe JH, Franklin B, Lavie CJ (2014) Exercising for health and longevity versus peak performance: different regimens for different goals. Mayo Clin Proc 89(9):1171–1175. https://doi.org/10.1016/j.mayocp.2014.07.007
3. Kreher JB (2016) Diagnosis and prevention of overtraining syndrome: an opinion on education strategies. Open Access J Sports Med 7:115–122. https://doi.org/10.2147/OAJSM.S91657
4. Meeusen R, Duclos M, Foster C et al (2013) Prevention, diagnosis, and treatment of the overtraining syndrome: joint consensus statement of the European college of sport science and the american college of sports medicine. Med Sci Sports Exerc 45(1):186–205. https://doi.org/10.1249/MSS.0b013e318279a10a
5. Fry RW, Morton AR, Keast D (1991) Overtraining in athletes. An update. Sports Med 12(1):32–65
6. Warren MP, Shantha S (2000) The female athlete. Baillieres Best Pract Res Clin Endocrinol Metab 14(1):37–53
7. Bø K, Stien R (1994) Needle EMG registration of striated urethral wall and pelvic floor muscle activity patterns during cough, valsalva, abdominal, hip adductor, and gluteal muscle contractions in nulliparous healthy females. Neurourol Urodyn 13(1):35–41
8. Sapsford RR, Hodges PW, Richardson CA et al (2001) Co-activation of the abdominal and pelvic floor muscles during voluntary exercises. Neurourol Urodyn 20(1):31–42
9. Kotarinos RK (2016) Musculoskeletal pelvic anatomy. In: Hoyte L, Damaser M (eds) Biomechanics of the Female Pelvic Floor. Elsevier Inc, pp 53–88
10. Gottschal JS, Kram R (2005) Ground reaction forces during downhill and uphill running. J Biomech 38(3):445–452
11. Leitner M, Moser H, Eichelberger P et al (2017) Evaluation of pelvic floor muscle activity during running in continent and incontinent women: an exploratory study. Neurourol Urodyn 36(6):1570–1576. https://doi.org/10.1002/nau.23151
12. Kruger JA, Dietz HP, Murphy BA (2007) Pelvic floor function in elite nulliparous athletes. Ultrasound Obstet Gynecol 30(1):81–85
13. Almeida MB, Barra AA, Saltiel F et al (2016) Urinary incontinence and other pelvic floor dysfunctions in female athletes in Brazil: a cross-sectional study. Scand J Med Sci Sports 26(9):1109–1116. https://doi.org/10.1111/sms.12546
14. Haylen BT, de Ridder D, Freeman RM et al (2010) An International Urogynecological Association (IUGA)/International Continence Society (ICS) joint report on the terminology for female pelvic floor dysfunction. Neurourol Urodyn 29(1):4–20. https://doi.org/10.1002/nau.20798

15. Nygaard IE, Thompson FL, Svengalis SL et al (1994) Urinary incontinence in elite nulliparous athletes. Obstet Gynecol 84(2):183–187
16. Eliasson K, Larsson T, Mattsson E (2002) Prevalence of stress incontinence in nulliparous elite trampolinists. Scand J Med Sci Sports 12(2):106–110
17. Da Roza T, Brandão S, Mascarenhas T et al (2015) Volume of training and the ranking level are associated with the leakage of urine in young female trampolinists. Clin J Sport Med 25(3):270–275. https://doi.org/10.1097/JSM.0000000000000012
18. Thyssen HH, Clevin L, Olesen S et al (2002) Urinary incontinence in elite female athletes and dancers. Int Urogynecol J Pelvic Floor Dysfunct 13(1):15–17
19. Jacome C, Oliveira D, Marques A et al (2011) Prevalence and impact of urinary incontinence among female athletes. Int J Gynaecol Obstet 114(1):60–63. https://doi.org/10.1016/j.ijgo.2011.02.004
20. Bø K, Borgen JS (2001) Prevalence of stress and urge urinary incontinence in elite athletes and controls. Med Sci Sports Exerc 33(11):1797–1802
21. Simeone C, Moroni A, Pettenò A et al (2010) Occurrence rates and predictors of lower urinary tract symptoms and incontinence in female athletes. Urologia 77(2):139–146
22. Vitton V, Baumstarck-Barrau K, Brardjanian S et al (2011) Impact of high-level sport practice on anal incontinence in a healthy young female population. J Womens Health (Larchmt) 20 (5):757–763. https://doi.org/10.1089/jwh.2010.2454
23. Milanesi R, Caregnato RC (2016) Intra-abdominal pressure: an integrative review. Einstein (Sao Paulo) 14(3):423–430. https://doi.org/10.1590/S1679-45082016RW3088
24. Bø K (2004) Urinary incontinence, pelvic floor dysfunction, exercise and sport. Sports Med 34(7):451–464
25. Da Roza T, Brandão S, Oliveira D et al (2015) Football practice and urinary incontinence: Relation between morphology, function and biomechanics. J Biomech 48(9):1587–1592. https://doi.org/10.1016/j.jbiomech.2015.03.013
26. Hay JG (1993) Citius, altius, longius (faster, higher, longer): the biomechanics of jumping for distance. J Biomech 26(S1):7–21
27. Nygaard IE, Glowacki C, Saltzman CL (1996) Relationship between foot flexibility and urinary incontinence in nulliparous varsity athletes. Obstet Gynecol 87(6):1049–1051
28. Luginbuehl H, Naeff R, Zahnd A et al (2016) Pelvic floor muscle electromyography during different running speeds: an exploratory and reliability study. Arch Gynecol Obstet 293(1): 117–124. https://doi.org/10.1007/s00404-015-3816-9
29. Caylet N, Fabbro-Peray P, Marès P et al (2006) Prevalence and occurrence of stress urinary incontinence in elite women athletes. Can J Urol 13(4):3174–3179
30. Bourcier AP, Alain P (2008) Incontinence during sport and fitness activities. In: Baessler K, Schüssler B, Burgio KL, Moore K, Stanton SL (eds) Pelvic floor re-education, Springer, pp 267–270
31. Ree ML, Nygaard I, Bø K (2007) Muscular fatigue in the pelvic floor muscles after strenuous physical activity. Acta Obstet Gynecol Scand 86(7):870–876
32. Poświata A, Socha T, Opara J (2014) Prevalence of stress urinary incontinence in elite female endurance athletes. J Hum Kinet 44:91–96. https://doi.org/10.2478/hukin-2014-0114
33. Córcoles MB, Sánchez SA, Bachs GJ et al (2008) Quality of life in patients with urinary incontinence. Actas Urol Esp 32(2):202–210
34. Xu D, Liu N, Qu H et al (2016) Relationships among symptom severity, coping styles, and quality of life in community-dwelling women with urinary incontinence: a multiple mediator model. Qual Life Res 25(1):223–232. https://doi.org/10.1007/s11136-015-1070-z
35. Hampel C, Artibani W, Espuña Pons M et al (2004) Understanding the burden of stress urinary incontinence in Europe: a qualitative review of the literature. Eur Urol 46(1):15–27
36. Nygaard I, Girts T, Fultz NH et al (2005) Is urinary incontinence a barrier to exercise in women? Obstet Gynecol 106(2):307–314. https://doi.org/10.1097/01.AOG.0000168455.39156.0f

37. Fozzatti C, Riccetto C, Herrmann V et al (2012) Prevalence study of stress urinary incontinence in women who perform high-impact exercises. Int Urogynecol J 23(12):1687–1691. https://doi.org/10.1007/s00192-012-1786-z
38. Salvatore S, Serati M, Laterza R et al (2009) The impact of urinary stress incontinence in young and middle-age women practising recreational sports activity: an epidemiological study. Br J Sports Med 43(14):1115–1118. https://doi.org/10.1136/bjsm.2008.049072
39. Margalith I, Gillon G, Gordon D (2004) Urinary incontinence in women under 65: quality of life, stress related to incontinence and patterns of seeking health care. Qual Life Res 13(8): 1381–1390
40. Neels H, Wyndaele JJ, Tjalma WA et al (2016) Knowledge of the pelvic floor in nulliparous women. J Phys Ther Sci 28(5):1524–1533. https://doi.org/10.1589/jpts.28.1524

Toward the Development of a Vaginal Finger-Cot Device for Measuring Pelvic Floor Muscles Strength

Augusto Gil Pascoal, Patrícia Silva and Fátima Sancho

Abstract Stress urinary incontinence (SUI) refers to the involuntary leakage of urine on effort or exertion, like sneezing or coughing and is a common problem in female population. Pelvic floor muscle (PFM) training has been demonstrated to be effective in the prevention and treatment of SUI. Vaginal squeeze pressure (VSP) is the most commonly used parameter on PFM strength, measured by vaginal palpation, by pressure transducers or by self-reported to the physiotherapist. Vaginal palpation is clinically relevant to evaluate PFM activation and to guide patient about correct PFM contraction. Vaginal palpation is questionable as a reliable method to quantify PFM strength nevertheless some detailed palpation scoring systems exist to quantify PFM strength. No single measurement instrument gives an overall view of PFM function. A combination of the advantages of vaginal palpation with a reliable measure of the VSP, in a single instrument, is a challenging purpose. This chapter reports the responsiveness and validity of a customized adjustable finger-cot (vaginal finger-cot device) with an embed strain gauge pressure transducer that allows to measured VSP in a time base. The finger-cot, developed by our research according to women's health physiotherapists' specifications, kept preserved the properties of vaginal palpation, minimizing the measurement artifacts inherent to intravaginal probe.

Keywords Urinary incontinence · Vaginal assessment · Pelvic floor muscles physiotherapy · Vaginal palpation

---

A. G. Pascoal (✉)
CIPER—LBMF, Faculty of Human Kinetics, University of Lisbon, Lisbon, Portugal
e-mail: gpascoal@fmh.ulisboa.pt

P. Silva
ESTeS Lisboa, Polytechnic Institute of Lisbon, Lisbon, Portugal

F. Sancho
The Alcoitão School of Health, Alcabideche, Portugal

S. Brandão et al. (eds.), *Women's Health and Biomechanics*,
Lecture Notes in Computational Vision and Biomechanics 29,
https://doi.org/10.1007/978-3-319-71574-2_13

Highlights

- No single measurement instrument gives an overall view of pelvic floor muscle function (PFM)
- An instrument that combines vaginal palpation with a reliable measure of the PFM strength is needed
- A customized adjustable vaginal finger-cot device (VFCD), with an embed strain gauge pressure transducer allowing to measure PFM strength in a time base seems adequate
- It preserves the properties of vaginal palpation while recording the intravaginal squeeze pressure during PFM contraction, minimizing the measurement artifacts inherent to intravaginal probe
- The VFCD can be guided by palpation and placed in direct contact with the levator ani, minimizing constraints related to co-contraction of the abdominal wall and diaphragm muscles.

1 Pelvic Floor Muscles Training and Stress Urinary Incontinence

The pelvic floor muscles (PFM) form a three-layer muscular surface expanding from the pubic symphysis toward the coccyx, along the frontal sidewalls of the ileum. The PFM comprise the pelvic diaphragm muscles, also referred as the deep layer (*pubococcygeus*, *puborectalis*, and *iliococcygeus*, together known as the *levator ani*); the urogenital diaphragm muscles or the superficial layer of the PFM (*ischiocavernosus*, *bulbospongiosus*, and *transversus perinei superficialis*, together known as the perineal muscles); and the urethral and anal sphincter muscles.

The dysfunction of the female PFM is the known cause of several clinical conditions, including urinary and fecal incontinence, pelvic organ prolapse with prolapse of the anterior vaginal wall (cystocele), posterior vaginal wall (rectocele), vaginal apex (enterocele), vaginal vault, pain, and sexual disorders [1].

Urinary incontinence is defined as "*the complaint of any involuntary leakage of urine*" [2], and is a common problem in the female population with prevalence rates varying between 10 and 55% in 15- to 64-year-old women [3]. The most frequent form of urinary incontinence in women is stress urinary incontinence (SUI), defined as "*involuntary leakage on effort or exertion, or on sneezing or coughing*" [4].

The PFM training has demonstrated to be effective in the prevention and treatment of SUI [5, 6]. Kegel [7] was the first to report the effect of specific strength training of the PFM on female urinary incontinence and pelvic organ prolapse. Many randomized controlled trials have demonstrated its effectiveness, and *consensus* exists in the sense that PFM training should be the first-line treatment for SUI and mixed urinary incontinence [3].

Voluntary contraction of the PFM before and during increase intra-abdominal pressure produces a lift of the pelvic and urogenital diaphragms in a cranial and forward/anterior direction while simultaneously generating a squeezing force around the urethra, vagina, and rectum [8]. Two main mechanisms seem to explain why and how PFM training may be effective in the prevention and treatment of SUI: (1) by modifying women's behavior based on conscious learning on how to contract the PFM before and during increased intra-abdominal pressure; (2) women are taught to perform regular strength training over time in order to build up "stiffness" and structural support of the pelvic floor [9].

2 Evaluation of the Pelvic Floor Function and Strength

The evaluation of PFM function and strength is part of the clinical assessment when there is a suspicion of SUI. It is important to define the best PFM training protocol to be used by patients and to provide biofeedback and motivate patients along the training program. During this process, the clinician (e.g., physiotherapist) seeks to maximize the elevation and occlusion action of the PFM. However, a co-contraction of other muscles such as the gluteals, hip adductors, and abdominals, which is normal in healthy women, must be avoided. In fact, none of these other muscles by their own isolated contractions can act as a structural support to the pelvic organs or produce a urethral closure pressure during increases in abdominal pressure. Measurement of the PFM muscle action becomes complicated due to the involvement of these muscles but also by the diaphragmatic form the PFM and its attachments to the endopelvic fascia and pelvic organs.

Currently, clinical assessment of PFM function is based mainly on the qualitative perception of the strength of contraction and endurance by vaginal palpation. Regarding the physiotherapy, its advantages include detecting changes in resting tone, identifying areas of atrophy and differentiating between the states of contraction and relaxation of the PFM. During vaginal examination, using the distal pad of the index finger, it is possible to palpate the perivaginal muscles. A definite bulging and lifting of the muscles are felt during contraction, and in both the contracted and relaxed states, areas of *levator ani* atrophy can be detected. Furthermore, a weak to strong contraction can be evaluated, including differentiation between the squeeze and the lift components of muscle contraction. In this context, the accuracy vaginal palpation depends on the skill and experience of the examiner [10].

2.1 Parameters and Instruments Used on Pelvic Floor Muscles Assessment

The PFM function can be evaluated by vaginal palpation; observation; needle, wire and surface electromyography (EMG); ultrasound; magnetic resonance imaging (MRI); and pressure measurements. These methods can be distinguished as those to measure the ability to contract (e.g., vaginal palpation, ultrasound) from those to quantify strength, such as the manual muscle test by vaginal palpation, manometry, and dynamometry.

In an attempt to obtain objective measurements of voluntary PFM strength several types of instruments were developed. Kegel [7] was the first to use a device to evaluate PFM contraction with an intravaginal probe connected to a manometer calibrated from 0 to 100 mmHg, the so-called perineometer. This device is supposed to measure the ability of the PFM to develop vaginal squeeze pressure; however, no data has been reported by Kegel [7] about responsiveness, reliability, or validity of this method. In fact, the term "perineometer" is somewhat misleading because the pressure-sensitive region of the probe of the manometer is not placed on the perineum, but in the vagina at the level of the *levator ani*.

The vaginal squeeze pressure is the most commonly used parameter to quantify PFM strength, particularly during maximum strength and resistance (endurance) measurements. For this measurement, the woman is asked to contract the PFM as hard as possible (maximum strength), sustain the contraction and repeat as many contractions as possible (resistance). Quantification of vaginal squeeze pressure can be done using manual muscle test by vaginal palpation, pressure transducers, or dynamometers. Currently, several types of vaginal pressure devices are available to measure vaginal squeeze pressure/force as a measure of the pelvic floor strength [10]. In newer types of apparatus, a specialized balloon catheter connected to a fiberoptic microtip and strain gauge pressure transducer has shown high responsiveness [11]. Measurements made by these instruments, similar to Kegel perineometer, suffer from three artifacts. First, the compliance of the balloon, which means that the device fails to measure PFM force isometrically and an eventual systematic bias exists due to the striated muscle's length–tension and force–velocity relationships. Second, the variability of the intravaginal squeeze pressure along the vagina [12]. Due to this artifact, there may be an inadequate differentiation between the effect of contraction of the *levator ani* muscles, the abdominal wall muscles, and gluteal muscle contraction [11]. Third, as these devices are placed inside the vagina none of them differentiates between the effect of an intra-abdominal pressure rise and a change in PFM contraction.

Instruments to measure vaginal squeeze pressure have been found to be reproducible [10], however, the registered squeeze pressure cannot be interpreted alone since every increase in intra-abdominal pressure will be registered by these instruments. To ensure a valid measurement, simultaneous observation of inward movement by vaginal palpation has been suggested [10].

2.2 *Vaginal Palpation to Assess Pelvic Floor Muscle Function*

Vaginal palpation was first described by Kegel [7] as a method to evaluate PFM function. The examiner places one finger in the distal one-third of the vagina and asks the woman to lift inward and squeeze around the finger. In the beginning, Kegel used vaginal palpation only to teach women how to contract their PFM and classified the contraction qualitatively as correct or not correct, not as a method to measure PFM strength.

More recently, vaginal palpation has been considered not sufficiently sensitive, reproducible or a valid method to measure PFM strength. Nevertheless, some detailed scoring systems to quantify PFM strength based on vaginal palpation exist with moderate to good intra- and inter-rater reproducibility [13].

These scoring systems, such as the digital measure [14] and the Oxford scale [15], subjectively quantify the PFM strength using scales such as "zero-to-five" or "fair-good-excellent". Their subjective and categorical natures have limited accuracy and reproducibility [13, 16], depending also on the examiner training and expertise. In fact, the examiner must have previously performed a wide range of PFM strength testing to support the subjective comparison that is implicit in these scoring systems.

Although their use is limited in the research setting, vaginal palpation is clinically relevant to evaluate PFM activation and to guide patient about the correct way to contract the PFM. By using vaginal palpation, it is possible to qualitatively assess the lift inward displacement of the PFM during their contraction and guide the use of more accurate instruments to measure the vaginal squeeze pressure. In fact, a correct contraction involves an observable inward movement of the perineum or the instrument, and training creates a downward movement [13].

3 Development of a Vaginal Finger-Cot to Measure PFM Strength

No single measurement instrument gives an overall view of PFM function. No data exist about responsiveness, reliability, and validity of the existed measurement tools, particularly about the synergic action of the PFM as a response to increased abdominal pressure, in real-life situations. A combination of the advantages of vaginal palpation with a reliable measure of the vaginal squeeze pressure, in a single instrument, is a challenging purpose. The best option to achieve this purpose seems to be the development of a clinical appropriated pressure/force measurement instrument, adjustable to the finger of the examiner, which can record the PFM strength (vaginal squeeze pressure) while preserving the vaginal palpation properties.

Recently a customized adjustable finger-cot with an embed strain gauge pressure transducer (Fig. 1) was developed by our team to measured vaginal squeeze pressure/force, in a time base, the vaginal finger-cot device (VFCD), according to the specifications proposed by women's health-skilled physiotherapists, in partnership with biosignals monitoring specialized company (Bio Signals Plux; www.biosignalsplux.com).

In order to test the clinical appropriateness of the VFCD, a convenient sample of nine women [age: 30 ± 8.7 years; nulliparous (N = 5) and primiparous (N = 4)] was recruited among a student population. The inclusion criterion was the ability to contract the PFM as assessed by vaginal palpation while the exclusion criteria were

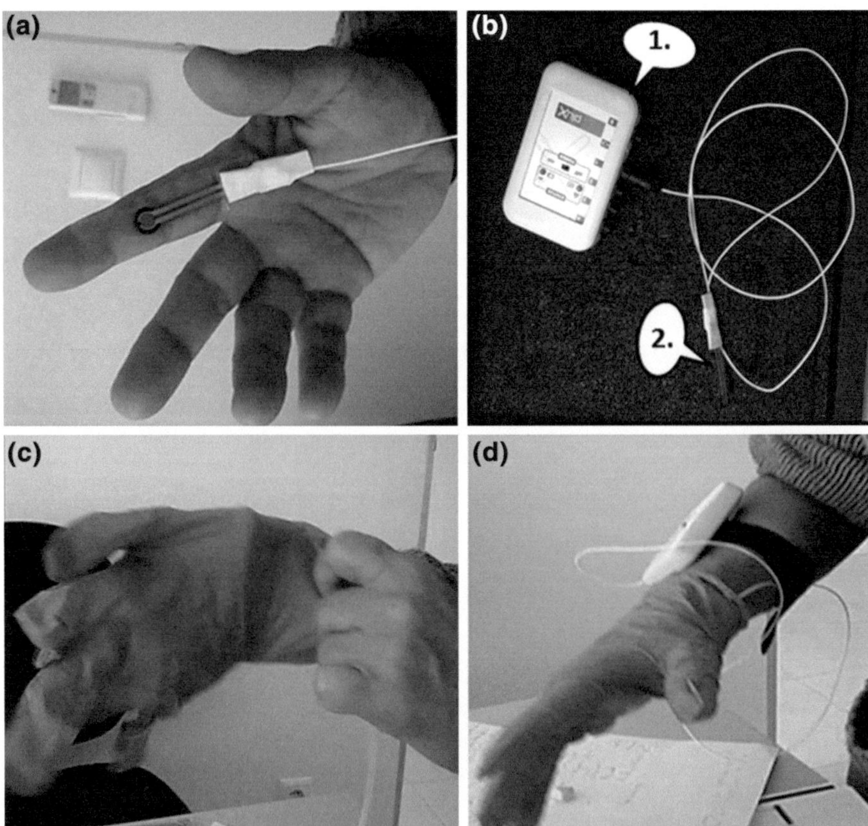

Fig. 1 The vaginal finger-cot device. **a** Pressure sensor placement; **b** Device components: [B1.] the wireless wearable clip signal hub unit; [B2.] the pressure sensor; **c** A disposable glove to cover the device; **d** Final configuration

pregnancy, pelvic surgery, pelvic or neurologic diseases, and menstruation on the day of assessment. Ethical approval for this study was obtained from the institutional Human Research Ethics Committee and subjects gave written informed consent. The PFM strength was assessed on each participant by two skilled physiotherapists, with the participants in lying supine position with a pillow under their head, straight knees and legs abducted. After thorough instruction in how to correctly contract PFM, a strong and fast contraction was asked against the finger of the examiner. The instruction used for each contraction was "squeeze and lift" or "tighten and pull up" the PFM. The sequence for muscle testing was as follows: three repetitions of maximum voluntary contractions lasting 3 s each with a 3-second rest period in between. No visible or palpable contraction of hip adductor, rectus abdominal or gluteal muscles was allowed. On each contraction, the vaginal squeeze strength was measured using the VFCD. Each of the physiotherapist examiners used one finger with the two distal phalanges inside the vaginal *introitus*. Measurements made by both examiners were mutually blinded. Calculation of intra-therapist reliability for maximum voluntary contraction was performed using the highest score of the three maximum voluntary contractions for each of the assessment methods. Mean of the mean of six contractions was used to estimate 95% confidence intervals (CI) for each examiner. During the assessment with VFCD three parameters were obtained from the sustained contraction (Fig. 2): (1) the peak pressure achieved during the contraction (PkFr); the rise time or time to achieve the peak, defined as time for the pressure signal to change from 10 to 90% of maximum intensity (TP); and the baseline pressure, defined as the pressure at rest with the patient relaxed (BP).

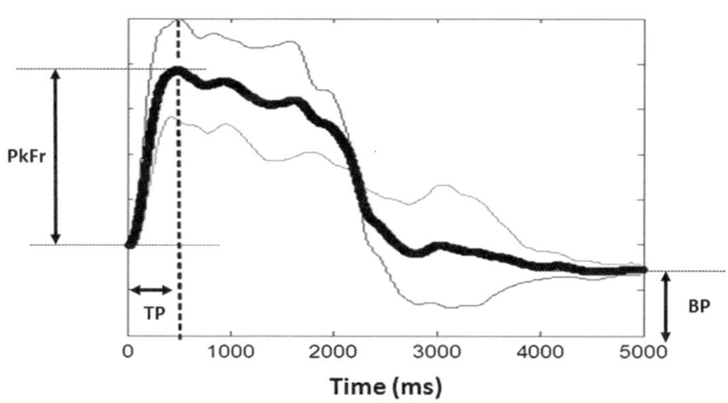

Fig. 2 Waveform of a sustained contraction: PkFr = Peak-to-Force, BP = Baseline Pressure, TP = Time-to-Peak

Intra-class correlation coefficients (ICC) were used to analyze the intra- and inter-observer reliability of the VFCD for measure peak force (PkFr) and time-to-peak force (Ti2PkFr) of the PFM. Reliability values were interpreted according to the scale by Altman [17], where <0.20 is considered poor, 0.20–0.40 fair, 0.41–0.60 moderate, 0.61–0.80 good, and 0.81–1.00 very good.

Two examples of the waveforms (before calibration) recorded by the VFCD are presented in Fig. 3. Notice the differences recorded on the amplitude of peak-to-force (PkFr) and on time-to-peak-to-force (Ti2PkFr). High intra- and inter-observer reliability were demonstrated for PkFr and Ti2PkFr (ICC = 0.82 and 0.87, respectively) for measurements of PFM strength with the VFCD.

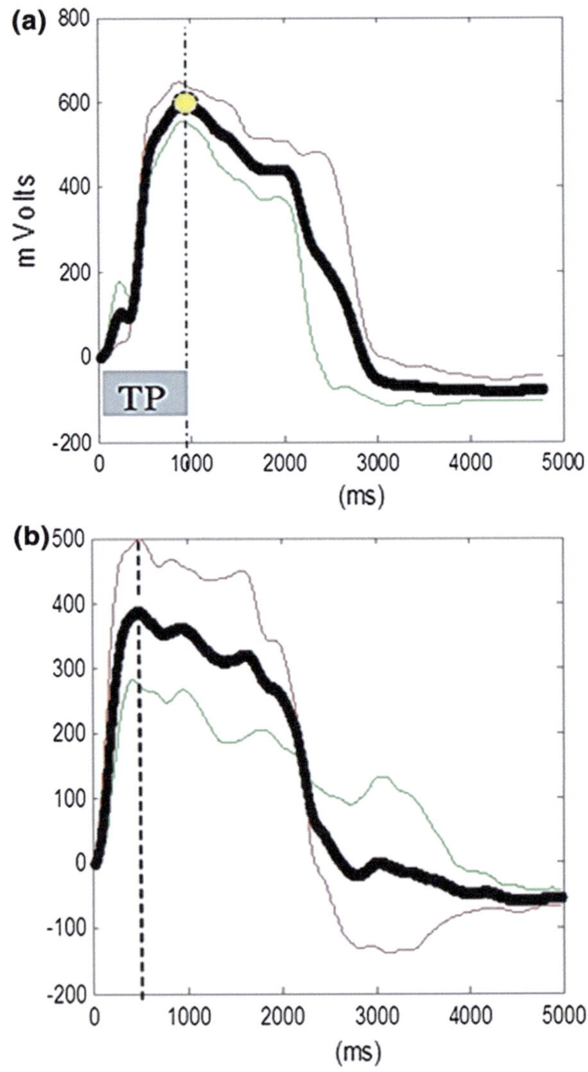

Fig. 3 Examples of two waveforms recorded by the vaginal finger-cot device. **a** Slow time-to-peak force (Ti2PkFr) and low vaginal squeeze pressure; **b** Fast TP and high vaginal squeeze pressure

4 Conclusions and Future Studies

Measurements of PFM function and strength are important for the analysis of the best PFM training protocol to be used by UI patients and may be an important tool to provide biofeedback and motivate patients during training programs. Vaginal palpation is commonly used in clinical practice to teach and evaluate PFM strength, and several grading systems have been developed. Vaginal devices including dynamometers and instrumented speculum were described for use in assessing the pressures generated by PFM actions [12].

The VFCD provides a valid measure of vaginal squeeze pressure either to use by a single or multiple physiotherapists on PFM strength assessment. This device was also developed to preserve the properties inherent to the vaginal palpation while recording the intravaginal squeeze pressure during PFM contraction. As the intravaginal squeeze pressure is known to vary along the vagina [7], guiding the pressure sensor placement along the vagina by palpation achieved by the VCFD ensures a more accurate PFM strength recording. The properties of vaginal palpation are kept preserved with the VFCD, which seems to minimize the measurement artifacts that involve the use of intravaginal probes. In fact, an accurate measure of the PFM strength is a challenge due to the frequent rise in the intra-abdominal pressure when women attempt to contract their *levator ani*. This is because the effort made to contract the muscles is accompanied by co-contraction of the abdominal wall musculature, and sometimes diaphragm muscle, which increasing intra-abdominal pressure. This measurement constraint is minimized by placing the VFCD in direct contact with the *levator ani*, guided by vaginal palpation.

This VFCD will allow an easier and more precise evaluation of the PFM contraction which is of great interest not only on grading the strength but also the ability to contract and at the same time giving a feedback to the patient. In the future, simple, affordable and easy to work devices like this must be present at every clinical practice where pelvic floor dysfunction conditions are treated. Evaluation, correct muscle contraction and biofeedback are the main keys to an optimum result of the condition.

References

1. Bump RC, Norton PA (1998) Epidemiology and natural history of pelvic floor dysfunction. Obstet Gynecol Clin North Am 25(4):723–746
2. Di Benedetto P (2004) Female urinary incontinence rehabilitation. Minerva Ginecol 56 (4):353–369
3. Bo K, Herbert RD (2013) There is not yet strong evidence that exercise regimens other than pelvic floor muscle training can reduce stress urinary incontinence in women: a systematic review. J Physiother 59(3):159–168. https://doi.org/10.1016/S1836-9553(13)70180-2
4. Bø K (2004) Urinary incontinence, pelvic floor dysfunction, exercise and sport. Sports Med 34(7):451–464

5. Bø K (2003) Pelvic floor muscle strength and response to pelvic floor muscle training for stress urinary incontinence. Neurourol Urodyn 22(7):654–658. https://doi.org/10.1002/nau.10153
6. Bø K (2004) Pelvic floor muscle training is effective in treatment of female stress urinary incontinence, but how does it work? Int Urogynecol J Pelvic Floor Dysfunct 15(2):76–84
7. Kegel AH (1948) Progressive resistance exercise in the functional restoration of the perineal muscles. Am J Obstet Gynecol 56(2):238–248
8. Bø K, Sherburn M, Allen T (2003) Transabdominal ultrasound measurement of pelvic floor muscle activity when activated directly or via a transversus abdominis muscle contraction. Neurourol Urodyn 22(6):582–588
9. Bø K (2006) Can pelvic floor muscle training prevent and treat pelvic organ prolapse? Acta Obstet Gynecol Scand 85(3):263–268
10. Bø K, Sherburn M (2005) Evaluation of female pelvic-floor muscle function and strength. Phys Ther 85(3):269–282
11. Ashton-Miller JA, Zielinski R, Miller JM et al (2014) Validity and reliability of an instrumented speculum designed to minimize the effect of intra-abdominal pressure on the measurement of pelvic floor muscle strength. Clin Biomech 29(10):1146–1150. https://doi.org/10.1016/j.clinbiomech.2014.09.011
12. Guaderrama NM, Nager CW, Liu J et al (2005) The vaginal pressure profile. Neurourol Urodyn 24(3):243–247
13. Bø K, Finckenhagen HB (2001) Vaginal palpation of pelvic floor muscle strength: inter-test reproducibility and comparison between palpation and vaginal squeeze pressure. Acta Obstet Gynecol Scand 80(10):883–887
14. Brink CA, Sampselle CM, Wells TJ et al (1989) A Digital test for pelvic muscle strength in older women with urinary incontinence. Nurs Res 38(4):196–199
15. Laycock J, Jerwood D (2001) Pelvic floor muscle assessment: the PERFECT scheme. Physiotherapy 87(12):631–642. https://doi.org/10.1016/S0031-9406(05)61108-X
16. Frawley HC, Galea MP, Phillips BA et al (2006) Reliability of pelvic floor muscle strength assessment using different test positions and tools. Neurourol Urodyn 25(3):236–242
17. Altman DG (2000) Statistics in medical journals: some recent trends. Stat Med 19(23):3275–3289

Physiotherapeutic Diagnostic Process for Female Urinary Incontinence

Bary Berghmans and Maura Seleme

Abstract The International Consultation on Incontinence (ICI) recommendations for initial management of female urinary incontinence, including pelvic physiotherapy, are based on presumed (medical) diagnosis, so indication, not (yet) diagnosis. This chapter demonstrates the role of a physiotherapeutic diagnostic process (PDP) for female urinary incontinence in clinical (research) practice. Furthermore, the technique and a practice protocol for the PDP including troubleshooting and interpretation are presented. Based on the PDP including the analysis, evaluation and based on that, the formulation of hypotheses, if indicated, optimal and adequate pelvic physiotherapy treatment can be selected.

Keywords Pelvic physiotherapy · Diagnostic process
Female urinary incontinence · ICF · Examination

B. Berghmans (✉)
Pelvic Care Center Maastricht, Maastricht University Medical Center, Maastricht
The Netherlands
e-mail: bary.berghmans@maastrichtuniversity.nl

M. Seleme
abafi-HOLLAND, Maastricht, The Netherlands

M. Seleme
Faculdade Inspirar, Curitiba, Brazil

S. Brandão et al. (eds.), *Women's Health and Biomechanics*,
Lecture Notes in Computational Vision and Biomechanics 29,
https://doi.org/10.1007/978-3-319-71574-2_14

Highlights

- The Physiotherapeutic Diagnostic Process (PDP) involves assessment of the nature and severity of female urinary incontinence in the context of whether its underlying disorders and/or any identified unfavorable prognostic factors are modifiable by physiotherapy
- The PDP is used to formulate a specific treatment plan, tailored to the individual's needs
- Using the 3 Incontinence Questionnaire (3IQ) test, the pelvic physiotherapist has a tool for rapid and valid distinction between the main types of urinary incontinence
- The validated PRAFAB questionnaire measures both the impact and the subjective severity of involuntary urine loss
- To evaluate pelvic floor muscle function, a structured assessment schedule based on evidence-based practice guidelines has been developed and incorporated in the PDP

1 Introduction

According to its latest published algorithms, the International Consultation on Incontinence (ICI) recommendations for initial management of Female Urinary Incontinence (FUI) including pelvic physiotherapy, are based on *presumed* (medical) diagnosis, so *indication, not (yet) diagnosis* [1]. This means that using the International Classification of Diseases [2] and mostly based on symptoms and signs, the diagnostic findings of the medical doctor end up as *indicative* for the specific health problem under investigation. As such, symptoms and signs by themselves do not refer to the underlying cause of the (pelvic floor) health problem. With this in mind, pelvic physiotherapy focuses on potential influence on this underlying cause and—if this is not possible/feasible—to develop a compensation strategy in order to minimalize the bother of the health problem. History taking and physical examination together lead to a hypothesis regarding the underlying cause and the possibility and extent of influence on it. To date, there is still no clear consensus about the role of pelvic physiotherapy in the assessment of FUI.

This chapter demonstrates the role of pelvic physiotherapy diagnostics for FUI in the clinical (research) practice. Furthermore, the technique and a practice protocol for the physiotherapeutic diagnostical process including troubleshooting and interpretation are presented.

During this process, which is used to formulate a specific treatment plan, the nature of the underlying disorder and UI severity (assessed on the basis of the International Classification of Functions, Disability and Health (ICF) [3] are examined in the context of whether the underlying disorders and/or any identified unfavorable prognostic factors are modifiable by physiotherapy.

1.1 Preparation

The complete physiotherapeutic diagnostic process will take place according to hygienic rules for pelvic physiotherapy, such as the "Guidelines for Hygienic Working in the Pelvic Floor Region" of the Dutch Association for Physical Therapy for Pelvic Floor Disorders and Pre- and Postnatal Healthcare [4].

First of all, especially in case there is no immediate medical attention, the pelvic physiotherapist screens for pathology requiring urgent medical attention. The pelvic physiotherapist should focus on recognizing typical patterns and identifying possible "red flags" (alarm signals), such as

- unexplained incontinence, pain while urinating, or blood loss
- signs of inflammation, infections, or general malaise
- fever
- (nocturnal) perspiring
- severe loss of weight
- loss of urine without the patients feeling the urge to empty their bladder and without pain

If the pattern indicates UI without any red flags, the further diagnostic process can be initiated immediately. The patient should be informed about the aim and execution of the physiotherapeutic diagnostic process, both verbally and in written, and should sign an informed consent. The patient is instructed to come to the clinic with an at least half full bladder (because of provocation tests during the physical examination) and not immediately to go to the toilet when entering the clinic.

The necessary information can be obtained by means of history taking, the patient's self-reports, (validated) questionnaires, micturition diaries and the therapist's own physical examination of the patient. We recommend using standardized questionnaires to collect data, such as the validated 3 Incontinence Questionnaire (3IQ) (Table 1) and the PRAFAB questionnaire (Table 3).

1.2 Pathophysiology Related to the PDP

Pelvic physiotherapists use the ICF dealing with the *consequences* of pelvic floor dysfunctions, such as FUI [3]. The pelvic physiotherapist will investigate the consequences of FUI on local level, i.e., disorders or impairments, such as weakness of the pelvic floor, rupture of connective tissue or pelvic floor muscle, etc.; on personal level, i.e., disabilities, limitation, such as not be able to lift a baby, or not able to stand up from a seat without involuntary urine loss; on sociocultural level, i.e., restriction in participation, such as not going anymore to the choir, church, marketplace without the fear of urine loss, bad odor, etc. (Table 2).

All these consequences are the basis for and the key elements of a following intervention, if indicated. Using the ICF and evaluating the findings of the so-called

Table 1 The 3 Incontinence Questionnaire test (Brown 2006)

1 During the last 3 months, have you leaked urine (even a small amount)?
Yes (please continue with questions 2 and 3)
No (questionnaire completed)
2 During the last 3 months, did you experience any involuntary loss of urine: (check all questions that apply)?
a. when you were performing some physical activity, such as coughing, sneezing, lifting, or exercise?
b. when you had the urge to empty your bladder, but you could not get to the toilet fast enough?
c. without physical activity and without a sense of urgency?
3 During the last 3 months, did you experience involuntary loss of urine most often: (check only one)
a. when you were performing some physical activity, such as coughing, sneezing, lifting, or exercise?
b. when you had the urge to empty your bladder, but you could not get to the toilet fast enough?
c. without physical activity and without a sense of urgency?
d. about equally as often with physical activity as with a sense of urgency?
4 Definitions of type of urinary incontinence are based on responses to question 3:
a. Most often with physical activity ➡ stress only or stress predominant urinary incontinence
b. Most often with the urgency to empty the bladder ➡ urgency only or urgency predominant (urinary) incontinence.
c. Without physical activity or sense of urgency ➡ other cause.
d. About equally with physical activity and sense of urgency ➡ mixed (urinary) incontinence.

Table 2 Definitions of the International Classification of Functioning Terms: Impairment, Disability, and Handicaps

Impairment: loss or abnormality of psychological, physiological, or anatomical structure or function at organ level. With respect to the classification of disorders in the storage and voiding of urine and feces, this means the impairment stress incontinence or detrusor overactivity.
Disability: restriction or loss of ability of a person to perform functions/activities in a normal manner. With respect to the classification of disabilities of voiding and stool, this means the disability involuntary loss of urine.
Restriction in participation: disadvantage due to impairment or disability that limits or prevents fulfillment of a normal role (depends on age, sex, sociocultural factors) for the person.

Source: WHO-Publication, *International Classification of Functioning, Disability and Health (ICF)*
World Health Organization, Geneva, Switzerland, 2001.

Physiotherapeutic Diagnostic Process (PDP), the pelvic physiotherapist formulates a preliminary hypothesis regarding (nature and extent of) the underlying problem and the extent of influence on its causing and maintaining factors. Based on this hypothesis the pelvic physiotherapist now is able to select the adequate treatment strategy, modalit(y)(ies) and its(their) parameters.

1.3 Establishing the Type of Incontinence

Goal-oriented and systematic history taking involve medical, paramedical, communicative, and attitudinal aspects. Every answer of the patient should lead the pelvic physiotherapist to the next question in a logic structure and sequence. Illness beliefs can have a favorable or adverse effect on the prognosis in terms of recovery and should be identified because beliefs may partly determine the type of intervention that can be used and can affect expectations of both the patient and the physiotherapist.

A recently developed short questionnaire for primary care, the 3IQ test, allows rapid, valid, and clearer distinction of the main types of urinary incontinence, viz. stress UI (SUI), urgency UUI, and mixed UI (MUI) during history taking [5]. The 3IQ test is a useful valid measurement instrument, which can establish the presence of SUI with a sensitivity of 0.86 (95% CI = 0.79–0.90) and a specificity of 0.60 (95% CI = 0.51–0.68), and can establish UUI with a sensitivity of 0.75 (95% CI = 0.68–0.81) and a specificity of 0.77 (95% CI = 0.69–0.84) [5].

1.4 Determining the Severity of the Health Problem

To assess the severity of the patient's health problem by identifying existence and extent of impairments, limitations of activities and restrictions of participation, the severity of the health problem is determined by the frequency and magnitude of the involuntary urine loss, the use of incontinence absorptive products and the consequences of UI for patient's everyday life, including work, sports and housekeeping activities, family life, social life, and sexuality.

1.5 PRAFAB Questionnaire

An instrument that assesses both urine loss and its impact on the patient is the validated PRAFAB questionnaire [6]. The PRAFAB questionnaire (Table 3) measures the severity of the urine loss in terms of the use of absorptive products (*PR*otection), the magnitude of the urine loss (*A*mount) and the number of times urine is lost (*F*requency). In addition, it records the impact of the urine loss in terms of more subjective aspects such as the way the patient adjusts to the urine loss in

Table 3 The PRAFAB questionnaire score (Hendriks 2007)

Protection
1. I never use protection for urine loss
2. I sometimes use protection, or I have to change my underwear because of urine loss
3. I normally use protection, or change my underwear several times a day because of urine loss
4. I always have to use protection because of urinary incontinence
Amount
1. The amount of urine loss is just a drop or less
2. Sometimes I loose a trickle
3. The loss of urine is so much that it wets noticeably my protection or clothes
4. The loss of urine is so much that my protection is soaked or leaks
Frequency **Involuntary loss of urine occurs:**
1. Once a week or less
2. More than once but less than three times a week
3. More than three times a week, but not everyday
4. Every day
Adjustment **Implications of urine loss:**
1. I am not hampered in my daily activities
2. I have stopped some activities, such as some sports and physically demanding activities
3. I have stopped most physical activities that caused involuntary loss of urine
4. I almost never go out
Body (or self) image
1. I am not bothered by my urine loss
2. I think urine loss is annoying and troublesome, but I am not greatly bothered by it
3. Urine loss makes me feel dirty
4. I am disgusted by myself because of my urinary incontinence

their everyday life (*A*djustment) and the consequences of the incontinence for the patient's self-image (*B*ody image). The PRAFAB questionnaire thus combines key objective and subjective aspects of the incontinence problem. The questionnaire measures two separate domains by means of a "leakage severity scale" (items *PR, A,* and *F*) and a "perceived impact scale" (items *A* and *B)* [7].

Total score:

* The PRAFAB questionnaire is validated in Dutch. Psychometric testing of the English version is not (yet) performed. Nevertheless, this questionnaire is provided in English to give the readers insight in the items and scoring system (min–max = 5–20 points; range 16 points)

1.6 *Identifying Etiological and Prognostic Factors*

Identification of all factors that may adversely affect the recovery process is needed in order to determine whether and how the health problem is amenable to modification by physiotherapy. This requires obtaining as much information as possible about:

- the nature of the underlying disorder, by identifying the presence of etiological factors for (the type of) UI, and
- any prognostic factors that influence the course of UI and the recovery process, viz.:
  - local factors that may adversely affect the recovery and adjustment/adaptation processes, whether relating to reduced resilience of or increased strain on the pelvic floor, or both;
  - general factors that may adversely affect the recovery process, that is, factors relating to the patient's poor physical and/or psychological condition

1.7 *Additional Testing for Type and Severity of Urinary Incontinence*

1.7.1 Micturition Diary

Also, a micturition diary is very useful. Such a diary (also sometimes called bladder or voiding diary) provides information about a number of variables relating to micturition, involuntary urine loss, and activities during which the involuntary urine loss takes place.

The following variables are systematically recorded, preferably covering at least 3 consecutive days that are representative of the patient's daily activity patterns, for instance, 2 working days and 1 weekend day [8].

- what and at what times the patient drinks fluids, and the amounts she consumes;
- the amount and timing of micturition;
- the level of urge(ncy) to empty her bladder;
- the moments when involuntary urine loss occurs and the amounts of urine loss

The micturition diary may support the assessment of both type and severity of urinary incontinence. A patient suffering from SUI usually has a normal voiding frequency (less or equal than 8 times in 24 h) and bladder volume, has mean micturitions between 200–400 cc/void, but with neither urgency nor nycturition. However, the patient complaints of losing small amounts of urine during exertion. A patient with UUI usually loses more urine (up to the complete content of the bladder) than a patient with SUI. On the other hand, the patient may lose less than 150 ml urine during micturition, suggesting a reduced functional capacity of the bladder.

1.8 Physical Examination

1.8.1 Inspection in Rest and During Movements

In healthy individuals, intra-abdominal pressure is automatically regulated by feed-forward control of the *transversus abdominis* muscle, with the diaphragm and the pelvic floor muscles (PFM) [9].

A strong relationship is described between lower back pain on one hand and UI and respiratory dysfunction on the other as the consequence of a limited ability to sufficiently integrate trunk muscle function in the regulation of posture and respiration as well as continence [10]. Good voiding and stool posture and relaxed pelvic floor and breathing create optimal conditions for emptying the bladder and intestines [11]. The pelvic physiotherapist should inspect the following (level of evidence 3) [10]:

- Patient's sitting and standing posture (urethral angle, anorectal angle, abdominal pressure, and toileting behavior)
- Respiration (breath holding and vocal behavior)
- Movements (mobility and tonicity of the spinal column, abdominal, and pelvic regions and movement patterns)
- Abdominal, buttock, and leg muscles (patients with fatigued PFM often show increased activity of other muscles)
- Local resilience can be reduced as a result of rupture or incisional scars by anterior or posterior vaginal wall defects or by uterine prolapse. Inspection can be used to identify signs of reduced pelvic floor resilience. The patient's specific resilience can be estimated from her general physical condition. Obesity is an unfavorable prognostic factor for recovery (level of evidence 3) and can be assessed using BMI measurement [12].

1.9 Physical Examination Techniques and Interpretation

To evaluate PFM function (Fig. 1), the following assessment schedule has been described in the Royal Dutch Society for Physiotherapy (KNGF) practice

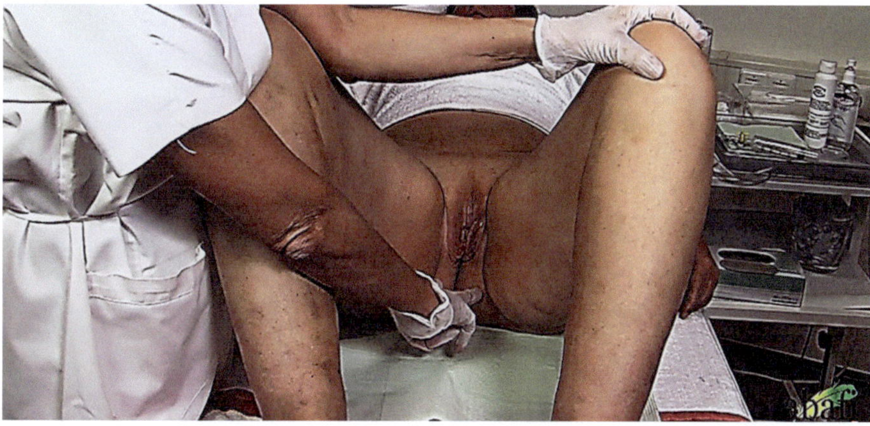

The optimal position to perform the physical exam is to ask for spreading the legs, one leg is supported between the body and elbow of the physiotherapist, the other leg by the PT's handof the same arm. In this way the PT can control whether there is any movement like more spreading or closing of the legs, any movement of the buttocks or the abdomen. This position is also very suitable to observe the perineum.

Fig. 1 Testing position of pelvic floor muscle functional assessment

guidelines "Dutch Guidelines for Physiotherapy in Patients with Stress Urinary Incontinence" (level of evidence 3) (Fig. 4) (eo) [13]:

- Assess whether the patient is able to voluntarily contract and relax the PFM, and evaluate the performance
- Assess the effectiveness of the voluntary contraction (Fig. 2) and relaxation of the PFM
- Assess the effectiveness of involuntary contraction of the PFM associated with a sudden increase in intra-abdominal pressure (forceful coughing) and subsequently during coughing after the patient has been instructed to contract their pelvic floor first
- Assess the effectiveness of involuntary relaxation of the PFM during straining
- Observe the voluntary contraction and relaxation of the PFM in relation to the abdominal muscles
- Establish any differences between the right and left side during an intravaginal digital palpation while the patient contracts and relaxes the PFM
- Quantify the strength, endurance (Fig. 2), and explosive strength of the PFM using manual muscle tests, such as vaginal or anal palpation [14] or using manometry [15] or dynamometry [16].

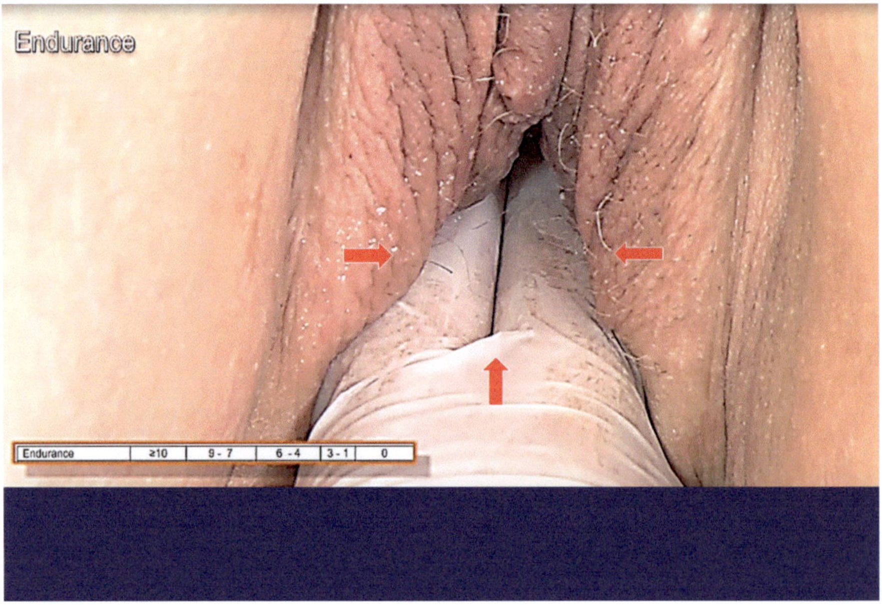

Fig. 2 Endurance of pelvic floor muscle contraction

2 Voluntary Control Over the Pelvic Floor Muscles

The pelvic physiotherapist should ascertain to what extent the patient has voluntary control, awareness, over her pelvic floor, as exercising or training the pelvic floor muscles can only be usefully done if the patient is able to voluntarily contract and relax her pelvic floor muscles. The "ability to contract" and "relax" can be assessed by means of clinical observation [14] vaginal or anal palpation (Fig. 3) [17] ultrasound [18], Magnetic Resonance Imaging (MRI) [19], and/or Electromyography (EMG) (Fig. 4) [8] although the correlation between EMG findings and muscle function in terms of strength, power, and endurance remains unclear [20].

Voluntary contraction of the PFM means that the patient is able to contract her pelvic muscles on demand. A contraction can be perceived as an encircling, elevating and tightening sensation around the palpating finger. The contraction may be "absent", "weak", "normal" or "strong" [21].

The strength of the PFM contraction is graded as [9, 26]: Absent, no contraction at all; Weak, i.e., weak contraction (PF lift not possible); Normal, i.e., moderate contraction (PF lift is possible); Strong, i.e., good contraction (PF lift is possible)

PFM relaxation should be tested after a contraction. Therefore, the investigator should always start with a contraction and then ask for relaxation. Voluntary relaxation of the PFM means that the patient is able to relax her pelvic muscles on demand after contracting them. The relaxation is perceived as the cessation of contraction. The PFM should at least return to their resting state. Voluntary "relaxation" can be rated as "absent", "partial", or "complete", "delayed" [21].

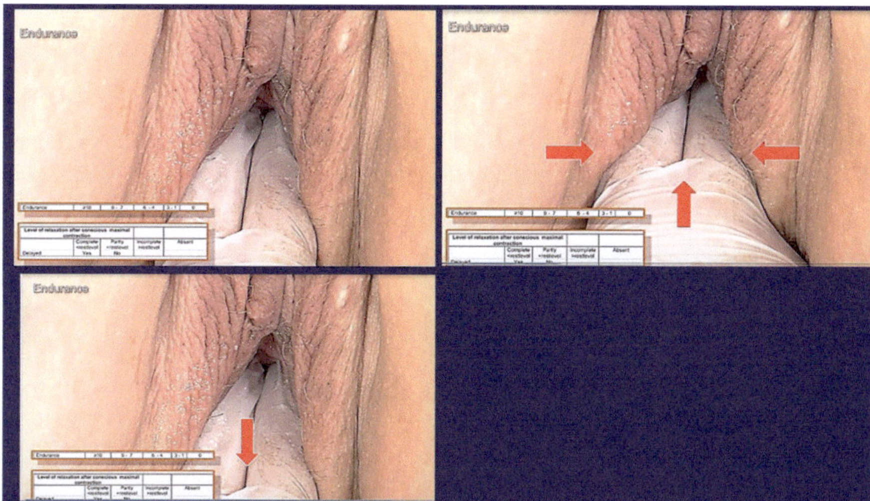

Levators closing	Strong	Normal	Weak	Absent
Symmetrc L/R	Yes	No	R > L	L > R
Level contraction	Strong	Normal	Weak	Absent

Fig. 3 PFM contraction in- and upward lifting of the palpating finger

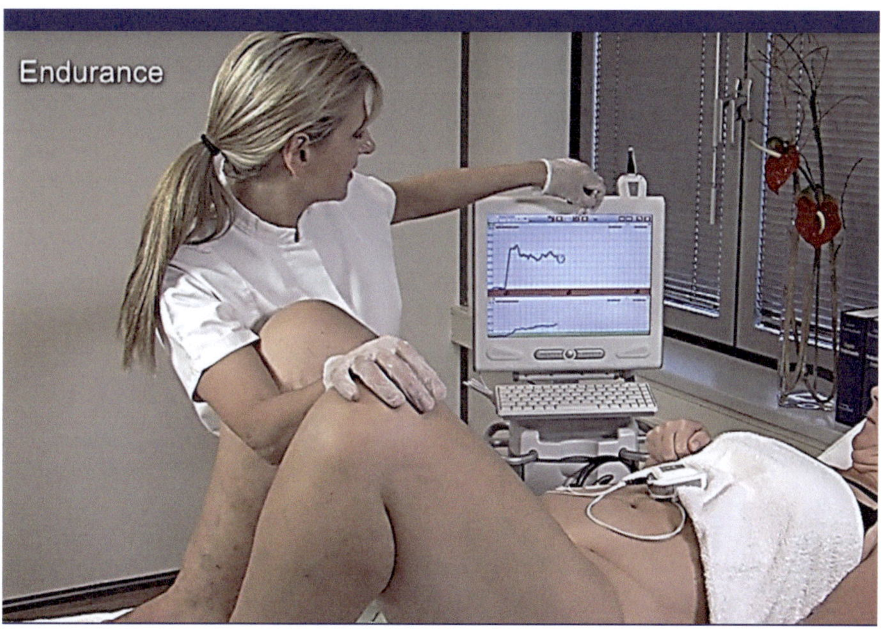

Fig. 4 EMG measurement during pelvic floor muscle functional assessment

3 Pelvic Floor Muscle Function

Normal PFM means:

- both voluntary and involuntary contraction and relaxation occur; the voluntary contraction is "normal" or "strong" and the voluntary relaxation is "complete";

Overactive PFM:

- no relaxation occurs; there is contraction even when relaxation is functionally needed, for instance during micturition or defecation;
- symptoms: micturition dysfunction, urine loss, obstructed defecation or dyspareunia;
- signs: absence of voluntary PFM relaxation;

Underactive pelvic floor muscles:

- no voluntary contraction occurs when it is appropriate;
- symptoms: urinary incontinence, anal incontinence;
- signs: pelvic organ prolapse and the absence of voluntary and involuntary contractions of the PFM;

Non-functioning PFM:

- no pelvic floor muscle activity palpable;
- symptoms: any symptom associated with a non-functioning pelvic floor;
- signs: any sign of a non-contracting, non-relaxing pelvic floor.

Information about verbal instructions during the PFM Functional Assessment (PFMFA) and how to interpret scores of the PFMFA can be found in Tables 4 and 5.

3.1 Additional Testing

More recently, an increasing number of pelvic physiotherapists assess pelvic floor function with perineal ultrasound. Dynamic evaluation of pelvic floor function includes position and elevation or descent of the bladder neck. Also, the *puborectalis* muscle at rest as well as pelvic floor pre-contraction, voluntary pelvic floor maximal and submaximal contractions, hold during respiration and sneezing or coughing, stabilization of the urethra, hold of bladder neck position during coughing or abdominal *maneuvres* can all be evaluated. However, although pelvic floor imaging using ultrasound becomes more and more popular, diagnostic ultrasound is reported to be well known for its operator-dependent nature and should only be used after appropriate and effective education [22].

Table 4 Pelvic Floor Muscle Functional Assessment (Slieker et al. 2009)

Patient(number) :
Researcher :
Date : Time:
Number of fingers : 1 2 Position:

INSPECTION during moving

Inw.mov. visible	Yes		No	Desc.
Cocontraction	No		Yes	
	RA/TrA	Dia-phragm	Adductors	Gluteal
Relaxation visible	Good	Delayed	Incomplete	Absent
Relaxation visible	Yes		No	

INSPECTION perineal movement during coughing and pushing

Coughing	Inw.	None	Desc.	
- in case inwards	before		During	after
Pushing	Desc		No	Inw.

PALPATION in rest

Pain	No	Yes	R	L	A	P
VAS	0-100					

PALPATION during moving

Conscious maximal contraction				
Urethral lift	Strong	Normal	Weak	Absent
Levators closing	Strong	Normal	Weak	Absent
Symmetrc L/R	Yes	No	R > L	L > R
Level contraction	Strong	Normal	Weak	Absent

Endurance (sec)	≥10	9 - 7	6 - 4	3 – 1	0
Explosive strength (n/15sec)	≥15	14 - 11	10 - 6	5 -1	0
Level of relaxation after conscious maximal contraction					
	Com-plete <restlevel	Partly =restlevel	Incom-plete >restlevel	Absent	
Delayed	Yes	No			
Unconscious contraction during coughing and pushing					
Coughing					
Reflexcontraction	Yes			No	
Descent perineum	Absent	Weak	Modest	Strong	
UI	No	Coughing	Pushing	Yes	
Flatus/FI	No	Coughing	Pushing	Yes	
Pushing					
Relaxation	Yes		No	Paradoxal	

CONCLUSION condition PFMF

Overactive	Normal	Coordination disorder	underac-tive	Non functional

Table 5 Interpretation scores Pelvic Floor Muscles Functional Assessment form (Slieker et al. 2009)

INSPECTION during movement

Inw.mov. visible	Yes	No	Desc..

Yes=a ventral and/or cranial movement of the perineum visible
No=no movement of the perineum visible
Desc.=the perineum moves caudal

Co-contraction visible	No	Yes

No=no contraction of surrounding muscles visible, small physiologic contraction m.transv.abd allowed, following or concurrent with a PFM contraction
Yes= contraction of surrounding muscles visible, mark which ones

Relaxation visible	Yes	No

Yes=after ending of contraction the perineum goes immediately back into rest position
No=the perineum does not go back, partly, delayed of hesitant back into the rest position

Coughing	Inw.	No	Desc.
- in case inward	before	during	After

Inw=cranial movement of the perineum visible
No= no caudal movement of the perineum visible
Yes= caudal movement of the perineum visible, mark moment of caudal movement: before, during or after coughing

Valsalva	Yes	No	Inw.

Desc.=dcaudal movement of the perineum visible
No= no caudal movement of the perineum visible
Inw=each inward movement of the perineum

PALPATION in rest

Pain	No	Yes	R	L	A	P
VAS	0-100					

No=no pain at all
Yes= all kinds of pain
R=right L=left A=anterior P=posterior, below mark VAS score (0= no pain, 100 is irresistible pain)

PALPATION while moving
Contraction

Urethral lift	Strong	Normal	Weak	Absent

Extent to which urethral lift is felt
Strong= maximal Knack, normal=….., weak…..

Levators closing	Strong	Normal	Weak	Absent

Extent to which levator closing is felt
Strong=palpating fingers are being closed, normal= fingers are pulled together but not completely, weak, small closing movement is felt, but inefficient

Symmetry	Yes	No	R>L	L>R

Yes= symmetry palpable
No= asymmetry palpable
R>L=right > left L>R=left > right

Level contraction	Strong	Normal	Weak	Absent

Strong=a powerful closing en cranio-ventral movement against firm resistance palpable
Normal=closing and cranio-ventral movement against light resistance palpable
Weak=short contraction, no palpable closing movement
Absent=no palpable response

Endurance (sec)	>10	9 - 7	6 - 4	3 - 1	0

>10 seconds
9 – 7
6 - 4
3 – 1
0

Explosive strength (n/15 sec)	>15	14 -11	10 --6	5--1	0

>15
14 – 11
10 – 6
5 – 1
0

Relaxation= relaxation following a contraction

Level relaxation	Complete < rest level	Partly = rest level	Incomplete > rest level	Absent
Delayed	Yes	No		

Complete=relaxation below restlevel
Partly=back to restlevel
Incomplete= relaxation above restlevel
Absent=no relaxation palpable
In case delayed; mark at relevant box as extra

Coughing

Reflexcontraction	Yes			No

Yes=during coughing reflex(inward)contraction of the PFM palpable No=no reflex(inward)contraction of the PFM palpable

Caudal movement perineum	Absent/inwards	Weak	Moderate	Strong

Absent=during coughing no caudal movement at all or no cranioventral movement
Present: level of occuring caudal movement
Weak= small caudal movement of max. 1 cm
Moderate= caudal movement 2-3 cm
Strong= caudal movement >3 cm

UI	No	coughing	Valsalva	Yes

No=dry, not a drop of UI
Yes=each way of UI

Flatus / FI	No	coughing	Valsalva	Yes

Flatus= loss of gas during coughing or Valsalva
FI= loss of feces during coughing or Valsalva

Valsalva

Unawared relaxation.	Yes		No	Paradoxal

Yes= palpable relaxation PFM
No=no palpable PFM relaxation
Paradox=contraction PFM

CONCLUSION condition PFM functions

Overactive	Normal	Coordination	Underactive	Nonfunctional

Coordination disorder=all combinations, such as:

- awareness of contraction, but no reflex,
- awareness of contraction, but no relaxation
- caudal movement PFM during inspection during contraction
- all asymmetric actions (R-L and A-P)
- weak contraction during contraction, partly relaxation during relaxation

A limitation of the different measurement methods, common to all clinic-based measurements of pelvic floor muscle function is that they are performed in the supine position or other standard positions. One should keep in mind that this might not reflect functional or usual activity of the pelvic floor during daily life activities as a response to increased abdominal pressure [17].

In healthy continent women, activation of the PFM before or during physical exertion seems to be an automatic response, so it is, in essence, an unconscious contraction [23]. This PFM "reflex" contraction is a fast feedforward loop and might precede the bladder pressure rise by 200–240 ms, something that might have been lost in FUI women [24]. Also, Bø has suggested that a well-timed, fast and strong PFM contraction may prevent urethral descent during intra-abdominal pressure rise [25]. So, PFM training (PFMT) is especially focused on adequate timing, strength improvement, adequate relaxation and coordination of the peri-urethral and the pelvic floor muscles. Appropriate treatment with PFMT should always include an assessment of PFM contraction and relaxation, because the effect of PFMT is dependent on whether the contractions and relaxations are performed correctly [25].

As timing of the PFM might be one of the most important elements this element needs to be included in the assessment. EMG biofeedback may serve this objective.

Biofeedback refers to a range of audiovisual techniques whereby information regarding "hidden" physiological processes, in this case, PFM contractions and relaxations, is displayed in a form understandable to the patient, to permit self-regulation of these events [26]. Usually, for assessment with EMG biofeedback, the motor unit activity (EMG) of the PFM at rest, during a maximal voluntary and involuntary (coughing and Valsalva) contraction (Pmax) and relaxation after Pmax are measured. The PFM functional assessment with wireless EMG biofeedback incorporates structured assessment in different positions, movements, and activities. Pre-contraction, timing, and coordination of the pelvic floor muscles are tested.

3.2 Analysis

The analysis stage involves the explicit decision whether "physical therapy" is the treatment indicated for the patient, based on the findings of the PDP, supplemented by the medical information that came with the referral. To this end, the pelvic physiotherapist has to answer the following questions: (1) Does the patient suffer from SUI, UUI, mixed UI and UI-related health problems? (2) What is the severity of the UI? (3) Are the pelvic floor muscles dysfunctional? (4) What is the cause of this dysfunction? (5) Are there currently any local prognostic factors that can adversely affect the recovery and/or adjustment processes, and can these local impeding factors be modified by physical therapy? (6) Are there currently any general prognostic factors that can adversely affect the recovery and/or adjustment processes, and can these general impeding factors be modified by physical therapy?

Based on the PDP including the analysis, evaluation and, based on that, the formulation of hypotheses, if indicated, pelvic physiotherapy treatment can start.

The various therapeutic interventions used by pelvic physiotherapists to treat female adult patients with UI include: (1) information and advice; (2) interventions to increase general physical condition; (3) intervention to increase the functional condition of the pelvic floor:

- practicing and controlling the pelvic floor functions (i.e., training the pelvic floor muscles to enable the patient to voluntarily contract and relax them, to increase their strength and endurance and to promote their involuntary contraction to support abdominal pressure increase);
- functional training;
- training with the help of feedback;
- training with the help of electrostimulation;
- training with the help of vaginal cones.

Altogether, the strategy how and when and to what extent to use these treatment modalities, the procedures (number of sessions, intensity, frequency), and tools (equipment, devices) form the physiotherapeutic therapeutic process.

4 Conclusions

To date, there is still no clear consensus about the role of pelvic physiotherapy in the *assessment* of FUI. Based on available evidence and relevant guidelines, a physiotherapeutic diagnostic process (PDP) for FUI in clinical (research) practice, including the technique, a practice protocol, troubleshooting, and interpretation of data are presented and discussed. The PDP involves assessment of the nature and severity of FUI in the context of whether its underlying disorders and/or any identified unfavorable prognostic factors are modifiable by pelvic physiotherapy. The PDP is used to formulate a specific treatment plan, tailored to the individual's needs.

References

1. Algorithm: initial management of urinary incontinence in women, p 1913. In: Abrams P, Cardozo L, Khoury S, Wein A (eds) Incontinence 5th international consultation on incontinence Paris, ICUD-EAU
2. WHO (1992) International Classification of Diseases ICD-10. World Health Organisation, Geneva, Switzerland
3. WHO (2001) International classification of functioning, disability and health. World Health Organisation, Geneva, Switzerland
4. Guidelines for hygienic working in the pelvic floor region. Dutch Association for Physical Therapy for Pelvic Floor Disorders and Pre- and Postnatal Healthcare (NVFB 2014)
5. Brown JS, Bradley CS, Subak LL et al (2006) The sensitivity and specificity of a simple test to distinguish between urge and stress urinary incontinence. Ann Intern Med 16;144(10):715–723
6. Hendriks EJ, Bernards AT, Berghmans BC et al (2007) The psychometric properties of the PRAFAB-questionnaire: a brief assessment questionnaire to evaluate severity of urinary incontinence in women. Neurourol Urodyn 26(7):998–1007. https://doi.org/10.1002/nau.20450
7. Hendriks EJ, Bernards AT, Staal JB et al (2008) Factorial validity and internal consistency of the PRAFAB questionnaire in women with stress urinary incontinence. BMC Urol 8:1. https://doi.org/10.1186/1471-2490-8-1
8. Staskin D et al (2013) Initial assessment of urinary incontinence in adult male and female patients (5A). In: Abrams P, Cardozo L, Khoury S, Wein A (eds) Incontinence 5th international consultation on incontinence, Paris, ICUD-EAU
9. Hodges PW, Sapsford R, Pengel LH (2007) Postural and respiratory functions of the pelvic floor muscles. Neurourol Urodyn 26(3):362–371
10. Smith MD, Russell A, Hodges PW (2006) Disorders of breathing and continence have a stronger association with back pain than obesity and physical activity. Aust J Physiother 52 (1):11–16
11. Devreese AM, Nuyens G, Staes F et al (2000) Do posture and straining influence urinary-flow parameters in normal women? Neurourol Urodyn 19(1):3–8
12. Townsend MK, Danforth KN, Rosner B et al (2007) Body mass index, weight gain, and incident urinary incontinence in middle-aged women. Obstet Gynecol 110(2 Pt 1):346–353
13. Bernards ATM, Berghmans BC, Slieker-Ten Hove MC et al (2014) Dutch Guidelines for physiotherapy in patients with stress urinary incontinence: an update. Int Urogynecol J 25 (2):171–179. https://doi.org/10.1007/s00192-013-2219-3

14. Slieker-ten Hove MC, Pool-Goudzwaard AL, Eijkemans MJ et al (2009) Face validity and reliability of the first digital assessment scheme of pelvic floor muscle function conform the new standardized terminology of the International Continence Society. Neurourol Urodyn 28 (4):295–300. https://doi.org/10.1002/nau.20659
15. Bø K, Sherburn M (2007) Vaginal squeeze pressure measurement. In: Bo K, Berghmans B, Møkved S, Kampen M van (eds) Evidence based physical therapy for the pelvic Floor. Bridging science and clinical practice. Butterworth Heinemann Elsevier, Edinburgh, pp 63–68
16. Dumoulin C, Gravel D, Bourbonnais D et al (2004) Reliability of dynamometric measurements of the pelvic floor musculature. Neurourol Urodyn 23(2):134–142
17. Bø K, Sherburn M (2005) Evaluation of female pelvic floor muscle function and strength. Phys Ther 85(3):269–282
18. Dietz HP, Jarvis SK, Vancaillie TG (2002) The assessment of levator muscle strength: a validation of three ultrasound techniques. Int Urogynecol J Pelvic Floor Dysfunct. 13(3):156–159
19. Stoker J, Halligan S, Bartram CI (2001) Pelvic floor imaging. Radiology 218(3):621–641
20. Vodusek DB (2007) Electromyography. In: Berghmans B, Møkved S, van Kampen M (eds) Evidence based physical therapy for the pelvic floor. Bridging science and clinical practice. Butterworth Heinemann Elsevier, Edinburgh, pp 56–63
21. Messelink B, Benson T, Berghmans B et al (2005) Standardization of terminology of pelvic floor muscle function and dysfunction: report from the pelvic floor clinical assessment group of the International Continence Society. Neurourol Urodyn 24(4):374–380. https://doi.org/10.1002/nau.20144
22. Dietz H (2007) Ultrasound in the assessment of pelvic floor muscle and pelvic organ descent. In: Evidence-based physical therapy for the pelvic floor: bridging science and clinical practice. In: Bø K, Berghmans B, Mørkved S, Van Kampen M (eds) London Churchill Livingstone Elsevier
23. Ashton-Miller JA, DeLancey JOL (2007) Functional anatomy of the female pelvic floor. In: Evidence-based physical therapy for the pelvic floor: bridging science and clinical practice. In: Bø K, Berghmans B, Mørkved S, Van Kampen M (eds) London Churchill Livingstone Elsevier
24. Constantinou CE, Govan DE (1982) Spatial distribution and timing of transmitted and reflexly generated urethral pressures in healthy women. J Urol 127(5):964–969
25. Bø K (1995) Pelvic floor muscle exercise for the treatment of stress urinary incontinence: an exercise physiology perspective. Int Urogynecol J Pelvic Floor Dysfunct 6(5):282–291
26. Moore K et al (2013) Adult conservative management. In: Abrams P, Cardozo L, Khoury S, Wein A (eds) Incontinence 5th international consultation on incontinence, Paris, ICUD-EAU

Psychological Morbidity, Sexual Satisfaction, Coping, and Quality of Life in Women with Urinary Incontinence in Rehabilitation Treatment

M. Graça Pereira, Susana Pedras and Cláudia Senra

Abstract This chapter focuses on the relationships between psychological morbidity, sexual satisfaction, coping, and quality of life in women with urinary incontinence in rehabilitation treatment, taking into consideration the women's perception of the severity of the urine loss and the type of urinary incontinence. Women that consider their incontinence symptoms as moderate reported better quality of life, and those with mild symptoms reported greater sexual satisfaction. Women with severe symptoms relied on religion and used more self-blame coping, expression of feelings, denial, and self-distraction strategies. Women with stress urinary incontinence used active coping while women with urge urinary incontinence used behavioral disinvestment, as coping strategies. Given the impact of urinary incontinence in women's lives, it would be important to develop health promotion strategies in order to help women cope and, as a result, improve their quality of life and the associated psychological symptoms.

Keywords Urinary incontinence · Psychological morbidity · Sexual satisfaction Coping · Quality of life

M. Graça Pereira (✉) · S. Pedras · C. Senra
Department of Applied Psychology, School of Psychology,
University of Minho, Braga, Portugal
e-mail: gracep@psi.uminho.pt

S. Brandão et al. (eds.), *Women's Health and Biomechanics*,
Lecture Notes in Computational Vision and Biomechanics 29,
https://doi.org/10.1007/978-3-319-71574-2_15

Highlights

- Worldwide, the prevalence of urinary incontinence is high
- Urinary incontinence is not exclusively a medical problem, and has repercussions on psychological well-being, sexual relationships, and quality of life
- The type of urinary incontinence and the severity of urine loss have a different impact on women
- Coping strategies are different according to the type of urinary incontinence and the severity of urine loss

1 Introduction

Until 1998, urinary incontinence (UI) was considered a symptom, but that changed and is now classified as a disease according to the International Classification of Diseases (ICD) [1]. The International Continence Society, defines the complaint regarding the involuntary loss of urine when associated with physical stress, such as effort, sneeze, or cough as stress urinary incontinence (SUI); when related to the loss, with or immediately preceded by urgency as urge urinary incontinence (UUI), and finally, as mixed urinary incontinence (MUI) when it occurs by urgency, effort, sneezing, or coughing [2].

Worldwide, more than 200 million of people are diagnosed with UI [3]. According to the International Continence Society, the estimated prevalence of women with UI in Portugal was 21.4% [3]. In a study conducted by Hunskaar et al., in 2000, with community-dwelling women who were not institutionalized [4], approximately 50% of women with UI were diagnosed with SUI, and this was the type of UI more prevalent in younger women [5]. Some other studies compared different UI types: the study of Coyne et al. [6] found that individuals with MUI refered more episodes of urine loss than with SUI and individuals with MUI and UUI reported more urine loss in a short period of time compared to SUI. In fact, women with SUI may have more control over the loss of urine by avoiding situations that increase abdominal pressure, whereas women with UUI have not [7, 8].

Regarding the risk factors, studies have reported that the prevalence of UI increases with age [9, 10], reaching a peak in middle age and growing continuously over the years [4]. Nevertheless, UI is not a direct consequence of the aging process [11], also affecting younger women, even in the absence of the traditional risk factors [12]. Pregnancy is also a risk factor, with an increase in the prevalence of SUI before and during pregnancy going from 5 up to 52% [13]. Data revealed that one-third of women experience SUI even 5 years after the first normal birth [14] and, therefore, vaginal birth is also listed as a risk factor for UI [1].

However, pregnancy and childbirth become less important with age [15]. Studies have found that the prevalence of SUI is high, particularly after menopause [1, 9], which may be related to estrogen deficiency [16]. Another factor that contributes to UI is obesity, due to an increase in intra-abdominal pressure [17]. Some behavioral habits, such as intense physical activity, tobacco, and caffeine consumption may also contribute to the frequency of urine loss [1, 18].

In fact, urinary incontinence may be considered a public health problem because of its magnitude, the costs required for both the individual and society, and the impact on quality of life [11]. Although UI is not a life-threatening disease [19], the loss of urine involves numerous repercussions on psychological morbidity and sexual relationships [20, 21], and quality of life [9] requiring the use of adaptative coping strategies.

2 Urinary Incontinence and Psychological Morbidity

According to Stach-Lempinen and colleagues [22], women with UI reported more depression, sadness, and loneliness when compared with the general population. Meade-D'Alisera et al. [23] compared women with and without UI and found that women with SUI were 3.6 times more likely to show depressive symptoms. Melville et al. [24] found that the prevalence of major depression in women with UI was 16%, and 7% for panic disorder. Besides depressive symptoms, women with UI, report anxiety symptoms which have a negative impact on QOL [22, 25]. Watson et al. [26] observed that women with UI and anxiety symptoms lost less urine than those without anxiety symptoms. The findings of depression were not straightforward associated with UI, with some discrepancies between positive [18] and negative associations [15, 27]. Regarding the prevalence of psychological morbidity in different types of UI, women with UUI or MUI also reported depressive symptoms [6]. According to the study of Lisicka and colleagues [25], depression was found in 7% of women with SUI, 3% with MUI and 2% with UUI. In comparison with women with SUI, the probability of women with UUI having depression was about 9.2 and 13.5% for MUI [24]. Similarly, Stach-Lempinen et al. [22] concluded that women suffering from UUI or MUI revealed significantly higher levels of severe depression when compared to women with SUI. Regarding anxiety, Lisicka and colleagues [25] found that 14% of women with SUI and MUI and 9% of women with UUI reported anxiety symptoms.

According to Melville et al. [24] and Vigod and Stewart [28], major depression significantly affect women's quality of life. In fact, women with UI showed low quality of life, more symptoms of depression, and the latter predicted low quality of life [22]. On the other hand, Meade-D'Alisera and colleagues [23] conducted a comparative study to assess the impact of depressive symptoms on sexual satisfaction and concluded that women with UI and depressive symptoms reported lower libido than women without UI. According to Vella and Cardozo [29], depression may lead to sexual dissatisfaction in women with UI since the loss of urine during intercourse may trigger repulsion regarding proximity and intimate contact [11].

3 Urinary Incontinence and Sexual Satisfaction

Sexuality plays a vital role in people's lives [30] being a complex issue strongly modulated by psychosocial situations [31]. The prevalence of UI in the general population during sexual relationships is about 2% [32]. In the study of Shaw [32], UI during sexual intercourse was described as an embarrassing and troublesome problem, showing a prevalence that ranged from 10 to 56%. In addition, sexual activity in women with UI seems to be significantly lower than in women without urine loss symptoms' [33, 34].

Regarding the impact of UI on sexual activity, Auge et al. [35] showed that women had difficulties in sexual intercourse, either by loss of urine, fear of interrupting the sexual activity to urinate, or simply by feeling shame toward the partner. In addition, UI may trigger lesions in the genital organs, pain, and discomfort, thus interfering with sexual response [11]. According to Vella and Cardozo [29], there are many reasons for destabilization in the sexual relationship, including post-surgery dyspareunia, urinary dermatitis, loss of libido, depression, and loss of urine during sexual intercourse. However, there is no consistency between studies, because there are few studies about the prevalence and impact of urine loss during sexual intercourse [32, 36].

Few studies have assessed the effect of different types of UI in sexual function. Urwitz-Lane and Özel [37] found that sexual function of women with UI did not differ according to the type of UI. Still, in the study of Lopes and Higa [12], MUI and SUI had a greater impact on women's sexual activity, while in the UUI, the most affected area was the social one. However, other authors report that SUI caused more discomfort during sexual intercourse than in social relations [13]. According to Temml et al. [33], UI interfered with marital and sexual life in 8 to 33% of women, and could lead to sexual dysfunction, which in turn, affects women's health and quality of life [30, 34], even more than sexual functioning [10, 35].

4 Urinary Incontinence and Coping

Coping strategies are very important in dealing with UI in order to maintain the identity and the perception of competence [38], since often UI is associated with lack of self-control [39]. One of the strategies is active coping that aims to initiate an action or efforts to control and limit the UI [40] such as going to the bathroom frequently or avoid physically demanding activities. Other strategies are used to minimize the effects of urine loss, such as using absorbents, hygiene products, and barrier creams [41] (to prevent infection and pain due to the contact of the genital organ with the urine) [11]. In addition to these strategies, women use more often dark and baggy clothes [42], to hide possible urine loss [43], perfumes [44], frequent bathing, change of underwear [21], liquid intake restriction, although some increase their liquid intake with the expectation that the UI symptoms will minimize

[45], weight reduction to decrease the loss of urine [46], support of religion [42], suspension of medication that may stimulate micturition, and also avoidance of social interaction [44]. Although some women seek medical help, many do not seek psychological support, even when their quality of life is affected [27] and the presence of urine loss is constant [10].

Denial prevents the adaptation to the UI. In the study of Bilgic et al. [47], 56% of women reported to have ignored, hide and declined to talk to someone when experiencing loss of urine for the first time. In this sense, the UI is often neglected [48] or even hide from the sexual partners [41].

Although there are no specific studies about other coping skills in UI, it is known that the use of social support as a coping strategy is associated with increased acceptance of the disease and well-being [49]. According to Carver et al. [50], this strategy is functional to the extent that provides tranquilization when the person is insecure and promotes problem-focused coping. Religion, in turn, is recognized as an important strategy for dealing with situations that involve stress [51], and women with UI often rely on religion to cope with this problem [42]. The expression of feelings may also be an adaptive response to deal with stressful situations [50]. Other adaptative coping strategies are positive reinterpretation, that requires the individual to interpret the stressful situation as an event with also a positive impact that allows personal growth and the use of humor, when women joke about the loss of urine and its impact [40]. The use of substances has a harmful effect on health and is a maladaptive coping strategy [52], as well as the use of self-distraction when women do not think or deal with the UI, leading to a maladaptive responses [53].

5 Urinary Incontinence and Quality of Life

According to the World Health Organization Quality of Life (WHOQOL) [54], quality of life depends on self-subjective perception toward their condition and treatment regarding the social, mental, and physical dimensions. In the study of Papanicolaou et al's. [7], over 80% of women reported UI symptoms being bothersome. Temml et al. [33] found that 66% of women reported that quality of life was affected by symptoms of UI, which had a greater impact on physical and social activities, trust and self-perception [7] than on daily life activities [13]. UI can lead to discomfort in social life by fear, shame, and humiliation during urine loss, which isolates women from friends and family [55], affecting the interpersonal relationships [56], and the participation in physical activities such as walking, running, playing, and dancing [12]. The UI also affects sleep, because may lead to fatigue and depression [57]. Vaart et al. [58] found that SUI, when compared with UUI, did not significantly affect quality of life. Similarly, from the three types of UI, the MUI and UUI were those that most affect women's lives [12, 59].

Although behavioral interventions are recommended as a first-line therapy in the treatment of UI [60], there is a paucity of research identifying effective psychological interventions or strategies in the improvement of quality of life in women with UI [61], as well as studies addressing the relationship between psychological morbidity, sexual satisfaction, coping strategies, and quality of life, in women with UI [62–65].

In the next section, we will describe a study done by our research team that focused on the relationships between sexual satisfaction, psychological morbidity, coping, and quality of life in women with UI that took into consideration the severity of the urine loss, according to the women's perception, and the type of UI.

6 Quality of Life in a Portuguese Sample of Women with UI

Eighty women diagnosed with UI, receiving rehabilitation treatment in a major Hospital in the North of Portugal participated in the study. Participants were adult woman diagnosed with UI with sexual activity, ranging from 27 to 80 years old, with a mean age of 45.59 years old (SD = 12.04). From the total sample, 71.3% of women had only 4 years of school education, 13.8% had 10 years of education, and 13.8% an university degree. Regarding the type of UI, 75% presented SUI, 17.5% UUI, and 7.5% MUI. 11.3% had undergone surgery for their UI. According to the severity of the symptoms of urine loss, 27.5% classified the UI as mild, 48.8% as moderate, and 23.8% as severe.

The UI severity and its impact on Quality of Life was assessed with the following instruments:

- *Incontinence Quality of Life (I-QOL)* [66] is composed of 22 items divided into three domains: avoidance and limiting behaviors (8 items), psychosocial impact (9 items), and social embarrassment (5 items). A higher score corresponds to a better quality of life. In this study, the Cronbach alpha was 0.95.
- *Satisfaction with Sexual Relationship Questionnaire (SSRQ)* [67] has 14 items organized into two domains: sexual relationship with eight items and confidence with six items. The confidence domain is divided into two subscales: self-esteem with four items and overall relationship with two items. A higher score implies higher sexual satisfaction. In this study, the Cronbach alpha was 0.97.
- *Hospital Anxiety and Depression Scales (HADS)* [68] is composed of 14 items grouped into two scales that measure anxiety and depression, with seven items each one and scored separately. A higher score corresponds to greater psychological morbidity. In this study, the Cronbach alpha is 0.78 for anxiety and 0.67 for depression.
- *Brief COPE* [69] has 28 items grouped into 14 domains: active coping, planning, positive reinterpretation, acceptance, humor, religion, seeking of emotional support, seeking of instrumental support, distraction, denial, expression of feelings, substance use, behavioral disengagement, and self-blame. Each one of these domains is composed of two items and scored separately. In this study, the Cronbach alpha ranged from 0.51 to 0.97, as in the original version. The subscales planning and seeking instrumental support were not used in the hypothesis testing due to low internal consistency (alpha below 0.70).

7 Relation Between Sexual Satisfaction, Coping, Psychological Morbidity, and Quality of Life

Results revealed a positive association between sexual satisfaction and quality of life ($r = 0.66$, $p < 0.01$) as well as between acceptance (coping) and quality of life ($r = 0.24$, $p < 0.05$) and a negative association between anxiety and quality of life ($r = -0.47$, $p < 0.01$) as well as between depression and quality of life ($r = -0.42$, $p < 0.01$). A negative relationship between the following coping strategies and quality of life was found: self-distraction ($r = -0.57$, $p < 0.01$), denial ($r = -0.69$, $p < 0.01$), self-blame ($r = -0.46$, $p < 0.01$), behavioral disinvestment ($r = -0.31$, $p < 0.01$), the expression of feelings ($r = -0.36$, $p < 0.01$), and religion ($r = -0.35$, $p < 0.01$). The results are shown in Table 1.

Higher quality of life was associated with greater sexual satisfaction. The section above is consistent with the general literature that shows a decrease in sexual activity in women with UI associated with a decreased quality of life [33, 35]. In fact, when the loss of urine occurs during sexual intercourse may lead to couple's disharmony, avoidance of sexual activity and embarrassment, impairing the quality of life [23, 29].

Higher quality of life was also associated with lower depressive and anxiety symptoms. The results corroborate the literature since depression as comorbidity is associated with decreased quality of life in women with UI [22, 24, 28], as well as anxiety since the uncertainty, frequency, and impact of urine loss often induces anxiety [70] and depression [71]. The impact of urinary incontinence on women's quality of life may also contribute to increased psychological morbidity [10].

Concerning the relationship between quality of life and coping strategies, the study of Xavier et al. [72] considers that quality of life depends on how women deal with disease and physical limitations and perceive their health condition. The results of this study found that a higher quality of life was associated with a greater acceptance of the UI. Although the literature is scarce to support this relationship, the study of Carr, Gibson, and Robinson [73] found that when women consider the UI natural for their age and adapt their routines and behaviors to deal with the problem, the quality of life may not be compromised. According to Lume [74], acceptance is an essential coping strategy to adapt to a disease and Abott and colleagues [75] found that this specific strategy was associated with better quality of life. Results also showed that a higher quality of life was associated with low levels of self-distraction, denial, self-blame, behavioral disinvestment, expression of feelings, and religion. Some studies with different samples found that high distraction [75], high self-blame [76], religious coping with a pessimistic view of the world [77], and high behavioral disinvestment [78] were associated with worse quality of life. Regarding the expression of feelings, the studies are contradictory, reporting that, the expression of feelings is associated with better quality of life when there is a perception of support [79], but the activation of emotions regarding talking and thinking of a stressful topic may lead to less quality of life [50]. Lazarus and Folkman [80] reported that denial might harm the individual's health hindering the adaptation to a particular disease.

Table 1 Pearson correlations among sexual satisfaction, psychological morbidity, coping, and quality of life

	1	2	3	4	5	6	7	8	9	10	11	12	13	14	15	16
1	–															
2	0.17	–														
3	0.12	−0.80	–													
4	0.31*	0.25*	−0.09	–												
5	0.11	0.00	0.03	−0.31**	–											
6	0.36**	−0.07	−0.02	0.31**	−0.21	–										
7	0.41*	0.33**	0.21	0.00	0.38**	−0.11	–									
8	0.06	0.27*	−0.21	0.30**	−0.19	0.08	0.08	–								
9	−0.11	0.02	−0.03	−0.31**	0.27*	−0.44**	0.23*	−0.12	–							
10	0.26*	0.18	0.25*	−0.11	0.60**	−0.26*	0.53**	−0.11	0.35**	–						
11	0.11	0.22	0.26*	−0.04	0.53**	−0.33**	0.48**	−0.03	0.32**	0.60**	–					
12	−0.12	0.21	−0.10	−0.05	0.42**	−0.33**	0.14	−0.10	0.22*	0.18	0.35**	–				
13	0.18	−0.08	0.18	−0.14	0.28*	−0.31**	0.37**	−0.35**	0.38**	0.44**	0.31**	0.12	–			
14	0.07	−0.03	0.14	−0.34**	0.28*	−0.24**	0.33**	−0.23*	0.41**	0.41**	0.16	0.10	0.54**	–		
15	−0.07	−0.03	−0.34**	0.19	−0.32**	0.31**	−0.24*	0.21	−0.31**	−0.55**	−0.45**	−0.13	−0.58**	−0.42**	–	
16	−0.15	0.01	**−0.35****	0.19	**−0.46****	**0.24***	**−0.36****	0.17	**−0.31****	**−0.69****	**−0.57****	−0.08	**−0.47****	**−0.42****	**0.66****	–
M	3.88	1.85	2.54	3.59	1.51	4.30	2.88	1.68	1.31	1.76	1.81	0.25	11.86	5.61	40.81	73.59
SD	1.26	1.74	2.27	1.41	1.59	1.02	1.76	1.88	1.45	1.72	1.94	0.93	4.60	3.35	14.21	19.10

**$p < 0.01$; *$p < 0.05$

1. Active Coping; 2. Seeking of Emotional Support; 3. Religion; 4. Positive Reinterpretation; 5. Self-blame; 6. Acceptance; 7. Expression of Feelings; 8. Humor; 9. Behavioral Disinvestment; 10. Denial; 11. Self-distraction; 12. Substance use; 13. Anxiety; 14. Depression; 15. Sexual Satisfaction; 16. Quality of Life

Women with higher quality of life were those who considered their symptoms of urine loss as mild or moderate. These results are in accordance with Papanicolaou et al.'s study [7], that found the severity of incontinence symptoms negatively associated with quality of life, i.e. women who reported symptoms as moderate/severe reported a more negative impact on their quality of life than those with mild symptoms. Other studies found similar results [27, 81]. Greater severity and discomfort due to UI symptoms were also associated with a negative impact on physical activity, trust, self-perception, and social activities [7] leading to a decrease on quality of life.

8 Differences in Sexual Satisfaction, Psychological Morbidity, Coping, and Quality of Life According to Women's Perceptions of the Severity of Urine Loss

The results revealed significant differences in quality of life according to the severity of the symptoms of urine loss ($\chi2$ (2) = 29.61, $p < 0.001$). Mann–Whitney tests with Bonferroni correction (0.05/3 = 0.017) found differences between the group of women with mild urine loss versus moderate; mild versus severe symptoms; and between moderate versus severe symptoms. Therefore, women who considered their symptoms of urine loss as moderate, when compared with those that considered as serious, reported higher quality of life.

Regarding sexual satisfaction, there were significant differences according to the severity of the symptoms of urine loss ($\chi2$ (2) = 7.38, $p < 0.05$). Mann–Whitney tests with Bonferroni correction (0.05/3 = 0.017) found differences between women with mild versus severe symptoms. Thus, women who considered their symptoms of urine loss as mild reported greater sexual satisfaction.

Concerning coping styles, differences were found on religion according to the severity of urine loss symptoms ($\chi2$ (2) = 7.67, $p < 0.05$) and Mann–Whitney tests with Bonferroni correction (0.05/3 = 0.017) showed that women with severe symptoms when compared to those with mild symptoms relied on religion; self-blame ($\chi2$ (2) = 6.81, $p < 0.05$) and Mann–Whitney tests with Bonferroni correction showed that women with self-blame coping reported severe urine loss compared to those with moderated and mild symptoms; expression of feeling ($\chi2$ (2) = 7.57, $p < 0.05$) and Mann–Whitney tests with Bonferroni correction showed that women with severe symptoms used more expression of feelings than those who considered the symptoms of urine loss as moderate; denial ($\chi2$ (2) = 14.64, $p < 0.01$) and Mann–Whitney tests with Bonferroni correction showed that women with severe symptoms of urine loss used more denial when compared with moderate or mild symptoms; self-distraction ($\chi2$ (2) = 15.60, $p < 0.001$) and Mann–Whitney tests with Bonferroni correction showed that women with severe symptoms used more self-distraction strategies when compared with mild and moderate symptoms. There were no significant differences in the remaining variables.

Table 2 Differences in sexual satisfaction, psychological morbidity, coping, and quality of life according to the severity of urine loss

	Mild (n = 22) Mean rank	Moderate (n = 39) Mean rank	Severe (n = 19) Mean rank	$\chi^2(2)$
Quality of life	57.07	42.23	17.76	**29.61*** **
Sexual satisfaction	48.64	41.49	29.05	**7.39***
Anxiety	34.48	42.13	44.13	2.15
Depression	37.77	41.08	42.47	0.47
Coping strategies				
Active coping	41.34	35.71	49.37	4.95
Emotional support	38.48	39.71	44.47	0.83
Religion	33.82	38.50	52.34	**7.67***
Positive reinterpretation	47.75	37.86	37.53	3.37
Self-blame	30.66	42.62	47.55	**6.81***
Acceptance	43.59	41.44	35.00	1.84
Expression of feelings	38.39	35.76	52.68	**7.57***
Humor	46.16	36.08	43.03	3.43
Behavioral disinvestment	31.70	44.83	41.79	5.24
Denial	30.09	38.60	56.45	**14.62****
Self-distraction	28.55	39.83	55.71	**15.60*****
Substance use	37.50	42.68	39.50	3.57

***$p < 0.001$; **$p < 0.01$; *$p < 0.05$

As Table 2 shows, women who considered their symptoms of urine loss as mild showed higher sexual satisfaction than those with severe symptoms. Norton and Brubaker [3] reported that women are more apprehensive about sexual activity because of the unpredictability of symptoms, since they fear the frequency of uncontrollable urine loss during sex, which may contribute to less sexual satisfaction.

Regarding coping strategies, the results revealed that women who considered the UI symptoms severe, practice more religious activities, when compared with those who considered the symptoms as mild. In the same sense, the women who feel guilty for having UI considered the symptoms of urine loss as severe. The literature is scarce in this area, however, the study of Koenig et al. [82] found that individuals with chronic conditions depend more on religious beliefs and practices in order to relieve stress, and better cope with health problems. Women with severe symptoms used more expression of feelings than those who considered the symptoms of urine loss as moderate. One may hypothesize that the perception of severity may lead to feelings of anxiety and uncontrollability, with a greater need for the expression of emotions. Future studies should corroborate this hypothesis Also, results showed more use of denial in women who perceive their symptoms as severe. This result is not in accordance with the literature since denial was reported in women who devalued the symptom, when the discomfort caused by the UI was low [83]. However, denial may be used as a way to reduce distress in situations that cause stress, and has been associated with better recovery in women submitted to surgery [84]. Therefore, in the case of UI, denial may be used when women feel they have no control over the situation. Future studies should also pursue this hypothesis. Finally, the results showed that women who considered the symptoms of urine loss as severe, used more self-distraction. One may hypothesize, that women rely more on activities to distract from the stressor that is interfering with their well-being [50].

9 Differences in Sexual Satisfaction, Psychological Morbidity, Coping, and Quality of Life According to the Type of UI

Differences were found on active coping strategies ($U = 263.50$, $p < 0.05$) and behavioral disinvestment ($U = 260.00$, $p < 0.05$) according to the type of UI. Women with SUI used active coping as a strategy, while women with UUI used behavioral disinvestment. There were no significant differences in the other variables.

Since SUI, when compared with other types of UI, is associated with the avoidance of physical exercises involving effort, instead of being related to the fear of not arriving in time to the bathroom [85], women may perceive they have more control and therefore use active coping strategies to avoid these activities. In fact, women deal better with the symptoms associated with SUI, because urine loss is predictable and minor, and therefore, active strategies may indeed decrease the frequency and the amount of urine loss [86] when compared to UUI explaining why, in this case, behavioral disinvestment might be more used.

10 Limitations of Studies Addressing Quality of Life in Women with UI

This study has some limitations, particularly the size of the sample, the cross-sectional design, and the fact that the sample included very few women with mixed incontinence. Also, all instruments were self-report and in the Brief COPE instrument, two subscales were not included in hypothesis testing due to low internal consistency.

Most studies carried in women with UI have also some limitations, related to the sample characteristics, particularly the small sample size, the inclusion of only married women, no involvement of partners, the lack of information about the underlying diseases, and patient selection (e.g., women seeking care in a hospital may have higher sumptom severity). Future studies should include bigger samples and address the impact of UI on professional activity and on physical exercise, particularly in young women, and therefore research in this area is needed.

11 General Conclusion

Given the impact of UI in women's lives, and in order to improve the quality of life and associated psychological symptoms, it would be important to develop health promotion strategies, including patient education regarding awareness of the risk factors, symptoms, and treatment of UI.

Behavioral interventions have demonstrated their efficacy in improving bladder control [61, 87] but more research is warranted. The study described in this chapter showed how psychological variables may be paramount in designing interventions that improve women's quality of life taking into consideration urine severity loss and type of UI.

References

1. ICD-10 (2010) International statistical classification of diseases and related health problems. World Health Organization
2. Abrams P, Cardozo L, Magnus F et al (2003) The standardization of terminology of lower urinary tract function: repot from the standardization sub-committee of the international continence society. Urology 61(1):37–49. https://doi.org/10.1016/S0090-4295(02)02243-4
3. Norton P, Brubaker L (2006) Urinary incontinence in women. Lancet 367(9504):57–67
4. Hunskaar S, Arnold EP, Burgio K et al (2000) Epidemiology and natural history of urinary incontinence. Int Urogynecol J Pelvic Floor Dysfunct. 11(5):301–319
5. Chadha G (2006) Urinary incontinence. Apollo Medicine. 3(1):48–53
6. Coyne KS, Zhou Z, Thompson C et al (2003) The impact on health-related quality of life of stress, urge and mixed urinary incontinence. BJU Int 92(7):731–735
7. Papanicolaou S, Hunskaar S, Lose G et al (2005) Assessment of bothersomeness and impact on quality of life of urinary incontinence in women in France, Germany, Spain and the UK. BJU Int. 96(6):831–838. https://doi.org/10.1111/j.1464-410X.2005.05722.x
8. Nitti VW (2001) The prevalence of urinary incontinence. Rev Urol. 3(1):2–6
9. Virkud A (2011) Management of stress urinary incontinence. Best Pract Res Clin Obstet Gynaecol. 25(2):205–216
10. Yip SK, Cardozo L (2007) Psychological morbidity and female urinary incontinence. Best Pract Res Clin Obstet Gynaecol. 21(2):321–329. https://doi.org/10.1016/j.bpobgyn.2006.12.002
11. Azevedo AA (2004) Incontinência urinária: impacto no feminino [Urinary incontinence: impact on female]. Formasau, Coimbra, p 2004
12. Villoroa R, Merino M, Hidalgo-Vega A et al (2016) Women with urinary incontinence in Spain: Health-related quality of life and the use of healthcare resources. Maturitas 94:52–57. https://doi.org/10.1016/j.maturitas.2016.08.010

13. Santos P, Mendonça D, Alves O et al (2006) Prevalência e impacte da incontinência urinária de stresse: antes e durante a gravidez [Prevalence and impact of stress urinary incontinence: before and during pregnancy]. Acta Med Port 19:349–356
14. Viktrup L, Lose G (2001) The risk of stress incontinence 5 years after first delivery. Am J Obstet Gynecol 185(1):82–87. https://doi.org/10.1067/mob.2001.114501
15. Schröder A et al (2010) Guidelines on Urinary Incontinence. European Association of Urology. Retrieved from: http://www.uroweb.org/gls/pdf/Urinary%20Incontinence%202010.pdf
16. Hendrix SL (2002) Urinary incontinence and menopause: an evidence-based treatment approach. Dis Mon 48(10):622–636
17. Subak LL, Johnson C, Whitcomb E et al (2002) Does weight loss improve incontinence in moderately obese women? Int Urogynecol J Pelvic Floor Dysfunct 13(1):40–43
18. Nygaard I, Turvey C, Burns TL et al (2003) Urinary incontinence and depression in middle-aged United States women. Obstet Gynecol 101(1):149–156
19. Lazzeri M, Novara G (2008) Urinary incontinence: ensure that the patient´s voice is heard. Eur Urol 53(6):1125–1128. https://doi.org/10.1016/j.eururo.2008.02.014
20. Kelleher C (2000) Quality of life and urinary incontinence. Baillieres Clin Obstet Gynaecol 14(2):363–379
21. Saleh N, Bener A, Khenyab N et al (2005) Prevalence, awareness and determinants of health care-seeking behavior for urinary incontinence in Qatari women: a neglected problem? Maturitas 50(1):58–65. https://doi.org/10.1016/j.maturitas.2004.04.003
22. Stach-Lempinen B, Hakala AL, Laippala P et al (2003) Severe depression determines quality of life in urinary incontinent women. Neurourol Urodyn 22(6):563–568
23. Meade-D´Alisera P, Merriweather T, Wentland M et al (2001) Depressive symptoms in women with urinary incontinence: a prospective study. Urol Nurs 21(6):397–399
24. Melville JL, Walker Edward, Katon Wayne et al (2002) Prevalence of comorbid psychiatric illness and its impact on symptom perception, quality of life, and functional status in women with urinary incontinence. Am J Obstet Gynecol 187(1):80–87. https://doi.org/10.1067/mob.2002.124839
25. Lisicka E, Dzigowski P, Radziszewski P et al (2010) Anxiety, depression and quality of life in women with urinary incontinence. Eur Urol 9(6):623
26. Watson AJS, Currie I, Curran S et al (2000) A prospective study examining the association between the symptoms of anxiety and depression and severity of urinary incontinence. Eur J Obstet Gynecol Reprod Biol 88:7–9. https://doi.org/10.1016/S0301-2115(99)00122-0
27. Minassian VA, Drutz HP, Al-Badr A (2003) Urinary incontinence as a worldwide problem. Int J Gynecol Obstet 82(3):327–338
28. Vigod SN, Stewart DE (2006) Major depression in female urinary incontinence. Psychosomatics 47(2):147–151
29. Vella M, Cardozo L (2005) Incontinence during sexual intercourse. Womens Health Med 2 (6):42–43
30. Xu Y, Song Y, Huang H (2011) Impact of the tension-free vaginal tape obturator procedure on sexual function in women with stress urinary incontinence. Int J Gynecol Obstet 112 (2):187–189. https://doi.org/10.1016/j.ijgo.2010.09.014
31. Dalpiaz O, Kerschbaumer A, Mitterberger M et al (2008) Female sexual dysfunction: a new urogynaecological research field. BJU Int 101(6):717–721. https://doi.org/10.1111/j.1464-410X.2007.07442.x
32. Shaw C (2002) A systematic review of the literature on the prevalence of sexual impairment in women with urinary incontinence and the prevalence of urinary leakage during sexual activity. Eur Urol 42(5):432–440
33. Temml C, Haidinger G, Schmidbauer J et al (200) Urinary Incontinence in both sexes: prevalence rates and impact on quality of life and sexual life. Neurourol Urodyn 19(3):259–271
34. Kim YH, Seo JT, Yoon H (2005) The effect of overactive bladder syndrome on the sexual quality of life in Korean young and middle aged women. Int J Impot Res 17(2):158–163

35. Auge AP (2006) Comparações entre os índices de qualidade de vida em mulheres com incontinência urinária submetidas ou não ao tratamento cirúrgico [Comparisons between the levels of quality of life in women with urinary incontinence submitted or not to surgical treatment]. Rev Bras Ginecol Obstet 28(6):352–357
36. Pais-Ribeiro JL, Raimundo A (2005) Satisfação sexual e percepção de saúde em mulheres com incontinência urinária [Sexual satisfaction and health perception in women with urinary incontinence]. Análise Psicológica 3(23):305–314
37. Urwitz-Lane R, Özel B (2006) Sexual function in women with urodynamic stress incontinence, detrusor overactivity, and mixed urinary incontinence. Am J Obstet Gynecol 195(6):1758–1761
38. Peake S, Manderson L (2003) The constraints of a normal life: the management of urinary incontinence by middle aged women. Women Health 37(3):37–51. https://doi.org/10.1300/J013v37n03_03
39. Brittain KR, Shaw C (2007) The social consequences of living with and dealing with incontinence – a carers perspective. Soc Sci Med 65(6):1274–1283
40. Pais-Ribeiro JL, Rodrigues AP (2004) Questões acerca do coping: a propósito do estudo de adaptação do brief COPE [Some questions about coping: the study of the Portuguese adaptation of the brief COPE]. Psicologia, Saúde & Doenças. 5(1):3–15
41. Newman DK (2001) Conservative management of urinary incontinence in women. Prim Care Update Ob Gyns 8(4):153–162
42. Beiji NK, Ozbas A, Aslan E et al (2010) Overview of the social impact of urinary incontinence with a focus on Turkish women. Urol Nurs 30(6):27–334
43. Diokno AC, Burgio K, Fultz NH et al (2004) Medical and self-care practices reported by women with urinary incontinence. Am J Manag Care 10(2):69–78
44. Borba AMC, Lelis MAS, Brêtas ACP (2008) Significado de ter incontinência urinária e ser incontinente na visão das mulheres [The meaning of having urinary incontinence and being incontinent on women's vision]. Texto Contexto Enferm 17(3):527–535
45. Elstad EA, Maserejian NN, McKinlay JB et al (2009) Fluid manipulation as a coping strategy among individuals with lower urinary tract symptoms: a mixed methods study. J Urol 181 (4):161
46. Anders K (2000) Coping strategies for women with urinary incontinence. Baillieres Best Pract Res Clin Obstet Gynaecol 14(2):355–361
47. Bilgic D, Kizilkaya Beji N, Ozbas A et al (2010) The coping and help seeking behaviours for management of urinary incontinence problem. Eur Urol Suppl 9(6):596
48. Sengezer T, Sunay D, Yildirim U et al (2010) Awareness, knowledge and help-seeking behaviors of women with urinary incontinence: primary results of a multicentrics study. Eur Urol Suppl 9(6):567–568. https://doi.org/10.1016/S1569-9056(10)61323-1
49. Reynolds P, Hurley S, Torres M et al (2000) Use of coping strategies and breast cancer survival: results from the black/white cancer survical study. Am J Epidemiol 152(10):940–949
50. Carver CS, Weintraub JK, Scheier MF (1989) Assessing coping strategies: a theoretically based approach. J Pers Soc Psychol 56(2):267–283
51. Krägeloh CU (2011) A systematic review of studies using the brief COPE: religious coping in factor analyses. Religions 2(3):216–246
52. Plaufcan MR, Wamboldt FS, Holm KE (2012) Behavioral and characterological self-blame in chronic obstructive pulmonary disease. J Psychosom Res 72(1):78–83. https://doi.org/10.1016/j.jpsychores.2011.10.004
53. Van den Akker OBA (2005) Coping, quality of life and psychological symptoms in three groups of sub-fertile women. Patient Educ Couns 57(2):183–189
54. The WHOQOL Group (1995) The world health organization quality of life assessment (WHOQOL): position paper from the world health organization. Soc Sci Med 41(10):1403–1409
55. Bogner HR, Gallo JJ, Sammel MD et al (2002) Urinary incontinence and psychological distress in community- dwelling older adults. J Am Geriatr Soc 50(3):489–495

56. Lee JJ (2005) The impact of urinary incontinence levels on the social lives of older Chinese in Hong Kong. Hallym Int J Aging 7(1):63–80. https://doi.org/10.2190/1775-31J1-4564-J352
57. Fultz N, Burgio K, Diokno AC et al (2003) Burden of stress urinary incontinence for community-dwelling women. Am J Obstet Gynecol 189(5):1275–1282
58. Vaart VD, de Leeuw JR, Roovers JP et al (2002) The effect of urinary incontinence and overactive bladder symptoms on quality of life in young women. BJU Int 90(6):544–549
59. Mallah F, Montazeri A, Ghanbari Z et al (2014) Effect of Urinary Incontinence on quality of life among Iranian women. J Family Reprod Health 8(1):13–19
60. Tran K, Levin RM, Mousa SA (2009) Behavioral intervention versus pharmacotherapy or their combinations in the management of overactive bladder dysfunction. Adv Urol 345324. https://doi.org/10.1155/2009/345324
61. Baker J, Costa D, Guarino JM et al (2014) Comparison of mindfulness-based stress reduction versus yoga on urinary urge incontinence: a randomized pilot study with 6-month and 1-year follow-up visits. Female Pelvic Med Reconstr Surg 20(3):141–146. https://doi.org/10.1097/SPV.0000000000000061
62. Lai HH, Rawal A, Shen B et al (2016) The relationship between anxiety and overactive bladder or urinary incontinence symptoms in the clinical population. Urology 98:50–57. https://doi.org/10.1016/j.urology.2016.07.013
63. Moore CK (2010) The impact of urinary incontinence and its treatment on female sexual function. Curr Urol Rep 11:299–303. https://doi.org/10.1007/s11934-010-0124-6
64. Çavan S, Yaman Ö, Orhan İ et al (2016) Prevalence of sexual dysfunction and urinary incontinence and associated risk factors in Turkish women. Eur J Obstet Gynecol Reprod Biol 203:303–308. https://doi.org/10.1016/j.ejogrb.2016.06.030
65. Bilgic D, Ozbas A, Aslan E et al (2015) Coping and help-seeking behaviors for management of urinary incontinence. LUTS. https://doi.org/10.1111/luts.12120
66. Patrick DL, Martin ML, Bushnell DM et al (1999) Quality of life of women with urinary incontinence: further development of the incontinence quality of life instrument (I-QOL). Urology. 53(1):71–76
67. Cappelleri JC, Althof SE, Siegel RL et al (2004) Development and validation of the self-esteem and relationship (SEAR) questionnaire in erectile dysfunction. Int J Impot Res 16 (1):30–38
68. Zigmond AS, Snaith RP (1983) The hospital anxiety and depression scale. Acta Psychiatr Scand 67:361–370
69. Carver CS (1977) You want to measure coping but your protocol´s too long: consider the brief COPE. Int J Behav Med 4(1):92–100
70. Frick A, Huang AJ, Van den Eeden SK et al (2009) Mixed urinary incontinence: greater impact on quality of life. J Urol 182(2):596–600. https://doi.org/10.1016/j.juro.2009.04.005
71. Aslan G, Köseoğlu H, Sadik O et al (2005) Sexual function in women with urinary incontinence. Int J Impot Res 17(3):248–251
72. Xavier FMF et al (2003) Elderly people's definition of quality of life. Rev Bras Psiquiatr 25 (1):31–39
73. Carr AJ, Gibson B, Robinson PG (2001) Measuring quality of life: is quality of life determined by expectations or experiences? BMJ 332(7296):1240–1243
74. Lume JR (1986) Comportamento humano em situação artificial da vida: um trabalho de investigação em hemodiálise renal [Human behavior in artificial situation of life: a research work on renal dialysis]. Faculdade de Medicina, Universidade do Porto; 1986
75. Abbott J, Hart A, Morton A et al (2008) Health-related quality of life in adults with cystic fibrosis: the role of coping. J Psychosom Res 64(2):149–157. https://doi.org/10.1016/j.jpsychores.2007.08.017
76. Klein DM, Turvey CL, Pies CJ (2007) Relationship of coping styles with quality of life and depressive symptoms in older heart failure patients. J Aging Health. 19(1):22–38. https://doi.org/10.1177/0898264306296398
77. Pargament K, Koenig H, Perez L (2000) The many methods of religious coping: development and initial validation of the RCOPE. J Clin Psychol 56(4):519–543

78. Taylor JL, Smith PJ, Babyak MA et al (2008) Coping and quality of life in patients awaiting lung transplantation. J Psychosom Res 65(1):71–79. https://doi.org/10.1016/j.jpsychores.2008.04.009
79. Stanton AL, Danoff-Burg S, Cameron CL et al (2000) Emotionally expressive coping predicts psychological and physical adjustment to breast cancer. J Consult Clin Psychol 68(5):875–882
80. Lazarus RS, Folkman S (1984) Stress, appraisal, and coping. Springer, New York
81. Di Gangi Herms AMR, Pinggera GM, De Jonge P et al (2003) Assessing health care needs and clinical outcome with urological case complexity: a study using INTERMED. Psychosomatics 44(3):196–203. https://doi.org/10.1176/appi.psy.44.3.196
82. Koenig HG, Larson DB, Larson SS (2001) Religion and coping with serious medical illness. Ann Pharmacother 35(3):352–359
83. Higa R, Lopes HMB (2007) Porque profissionais de enfermagem com incontinência urinária não buscam tratamento [Why nursing professionals with urinary incontinence do not seek treatment]. Rev Bras Enferm. 60(5):503–506
84. Wilson JF (1981) Behavioral preparation for surgery: benefit or harm? J Behav Med 4(1):79–102
85. Lagro-Janssen T, Smits A, Weel V (1992) Urinary incontinence in women and the effect on their lives. Scand J Prim Health Care 10(3):211–216
86. Coyne KS, Sexton CC, Irwin DE et al (2008) The impact of overactive bladder incontinence and other lower urinary tract symptoms on quality of life, work productivity, sexuality and emotional well-being in men and women: results from the EPIC study. BJU Int 101 (11):1388–1395. https://doi.org/10.1111/j.1464-410X.2008.07601.x
87. Tovian SM (1994) Urinary Incontinence: assessment, intervention, and the role of clinical health psychology in program development. J Clin Psychol Med Settings 1(4):339–361

Part V
Biomechanical Analysis of the Female Pelvic Floor

Searching for the Tissue Mechanical Properties in Pelvic Floor Dysfunction by Computational Modeling

Elisabete Silva, Sofia Brandão, Marco Parente, Teresa Mascarenhas and Renato Natal Jorge

Abstract The mechanical characteristics of the female pelvic floor are relevant to understand the pelvic disorders. These disorders can be associated with different risk factors, as hormonal changes, vaginal delivery, and obesity among others, and thus will affect their mechanical response. Urinary incontinence and pelvic organ prolapse are the main pelvic disorders and have been studied by using different methods, through experimental tensile using fresh female cadavers or tissues collected at the time of transvaginal hysterectomy and by applying imaging techniques, such as magnetic resonance imaging and ultrasound. But, the experimental analysis is important to determine biomechanical properties that can be used in pelvic modeling. The inverse Finite Element Analysis (FEA) allows estimating the in vivo biomechanical properties of the pelvic floor muscles for a specific subject—woman without pathology, with stress urinary incontinence, or pelvic organ prolapse–using input information acquired noninvasively.

Keywords Biomechanical tissue properties · Pelvic floor dysfunctions
Inverse finite element analysis

E. Silva (✉) · S. Brandão · M. Parente · R. Natal Jorge
LAETA, INEGI, Faculty of Engineering, University of Porto, Porto, Portugal
e-mail: silva.elisabete3@gmail.com

S. Brandão
Department of Radiology, Centro Hospitalar de São João—EPE, Faculty of Medicine, University of Porto, Porto, Portugal

T. Mascarenhas
Department of Gynecology and Obstetrics, Centro Hospitalar de São João - EPE, Faculty of Medicine, University of Porto, Porto, Portugal

S. Brandão et al. (eds.), *Women's Health and Biomechanics*,
Lecture Notes in Computational Vision and Biomechanics 29,
https://doi.org/10.1007/978-3-319-71574-2_16

Highlights

- Computational modeling helps to understand the disorders associated with female pelvic floor tissues.
- The inverse FEA coupled with MRI allows obtaining in vivo biomechanical properties of the PFM in a noninvasive manner.
- The biomechanical properties of the female PFM are relevant when explaining pelvic disorders since these disorders may result directly from changes in those properties.

1 Introduction

The female pelvic floor is a soft tissue support structure that includes fascia, ligaments, and muscles of the *anal* and *urogenital* regions that extends from the *symphysis pubis* to the coccyx, and comprises the *levator hiatus* for the passage of three orifices—the urethra, vagina, and rectum (Fig. 1) [1]. These structures are associated with different disorders that affect the quality of life of many women, and can be associated with changes in the biomechanical properties of the muscles, ligaments, and fascia [2]. In the case of the stress urinary incontinence (SUI), the involuntary urinary leakage associated with the increase in intra-abdominal pressure (IAP) is often associated with insufficient ligamentous support of the urethra, or weakness of the pelvic floor muscles (PFM). Also, pelvic organ prolapse (POP)—the descent of one or more pelvic structures from the normal anatomical location toward or through the *levator hiatus* opening—is related to the weakness or direct injuries of the PFM associated with different factors. The most common risk factors responsible for these changes are the hormonal changes due to age or menopause, obesity, ethnicity, vaginal surgeries, and vaginal delivery [1].

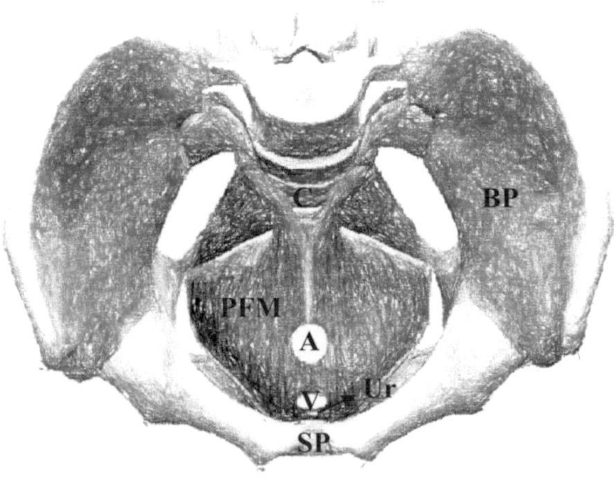

Fig. 1 Female pelvic floor anatomy. (A—anus; BP—bony pelvis; C—coccyx; PFM—pelvic floor muscles; SP—*symphysis pubis*; S—sacrum; Ur—urethra; V—vagina)

The *levator ani* is a major component of the PFM, and includes the *puborectalis*, *pubococcygeus*—which together are often called the *pubovisceralis* muscle (PVM) —and the *iliococcygeus* muscles. The PVM forms a sling around the rectum, vagina, and urethra that promotes the closure of the *levator hiatus*. The impairment of the PVM is a major ingredient to develop IU and/or POP [1].

Magnetic Resonance (MR) images help to exhibit the (lack of) support to the pelvic organs and the widening of the *levator hiatus* due to muscle defects or ligaments rupture [3]. By performing additional sagittal dynamic images during the Valsalva maneuver and coughing, one can evaluate the effect of the increase in the IAP over the pelvic viscera, and the counteracting (in)voluntary action of the muscles to suspend the organs [4–6]. Furthermore, it is already possible to apply this technique to the in vivo analysis of the mechanical properties of the soft tissues by performing MR elastography—it uses mechanical waves made by hardware probes to quantify tissue response. The stiffness and strain can be obtained by analyzing the induced motion, images, and quantitative measures of the tissue mechanical properties [7, 8]. MR elastography has been applied to upper and lower limb muscles during relaxation and contraction [9, 10]. These techniques are important for biomechanical analysis and validation of simulation results, which are noninvasive tools. However, they are still not available in all institutions and require additional hardware and software [11].

In this way, most of the literature still focuses on ex vivo studies of the pelvic ligaments and fascia [12–14] using mechanical tests that allow to understand their biomechanical behavior [12, 14]. Therefore, few studies have focused on the PFM, due to the ethical challenges in obtaining samples [2, 15], but ex vivo studies or animal models were used to investigate how risk factors (i.e., menopause and delivery) affect pelvic soft tissues biomechanical properties [2, 15]. But also the computer models were set to study their behavior during vaginal delivery or defecation [16–19].

Numerical simulations of the mechanical behavior of the PFM—based on the finite element method (FEM)—use in vitro biomechanical properties obtained from experimental studies with both normal and pathological specimens [12, 20]. In order to obtain subject-specific PFM geometrical and material properties, Silva et al. [21] implemented an inverse Finite Element Analysis (FEA) to estimate the most suited material constants for the PFM. This methodology allowed estimating the in vivo biomechanical properties for different conditions—without pathology, with SUI, and with POP—in a noninvasive way through MR images.

2 Mechanical Characterization of Female Pelvic Floor Disorders

Pelvic floor disorders (e.g., urinary incontinence (UI) and POP) have been previously studied by different authors, who concluded that weakness or direct injury to the PFM and the pudendal nerve may occur during vaginal delivery [22, 23]. In this sense, the biomechanical analysis of pelvic floor tissues is important to understand the different pelvic disorders to be able to improve clinical outcome by better understanding the effect of the decreased elasticity of the tissues, which often causes inability to maintain the normal position of the pelvic organs. These disorders may result from changes in the mechanical properties of the supportive structures that occur from impairment of muscles or ligaments, or alteration in the stiffness of the pelvic fascia associated with the previously referred risk factors [2].

In the case of UI, abnormalities in the pelvic ligaments and fascia can be associated with decreased total collagen and elastin [24]. Furthermore, Yip et al. [25] and Brandão et al. [19] also showed—through biomechanical modeing—that the effect of structural degradation of the support structures such as the ligaments and fascia is a major ingredient to develop SUI and cystocele. In addition, Delancey et al. [26] stated that SUI is also associated with weakness of the *levator ani* complex and decreased strength of the urethral sphincter, not always in the presence of direct injury to the PFM. In this sense, MR and ultrasound imaging allow exploring the functional analysis of the PFM, and these techniques can be used to compare continent and women diagnosed with SUI or POP [27, 28]. Although the biomechanical assessment is important, the MR imaging shows anatomic information on the status of the pelvic floor due to the high soft tissue contrast and the availability of acquiring functional images [28].

POP has been previously associated with muscle defects, but there is also evidence that women with thinner muscles may be more prone to organ descent [29]. Previous works studied the vaginal tissue in women with and without POP through experimental tension tests, using tissues collected at the time of transvaginal hysterectomy and of fresh female cadavers without prolapse [30–32], and they evidenced that those tissues present increased stiffness. Accordingly, mechanical testing provides important information about physical behavior of a specific soft tissue [33], and these ex vivo analyses are important to determining the mechanical properties [34] to provide input for pelvic modeling using the FEM [35].

In conclusion, the uniaxial tension tests show that pelvic ligaments have a nonlinear mechanical behavior, i.e., a nonlinear relationship between force and displacement, but they do not reflect the natural conditions for in vivo muscle behavior in the pelvis, and therefore computational analysis may help in simulating those in vivo relations and behavior.

3 Biomechanical Models of the Pelvic Floor Muscles

The PFM numerical models are important to simulate the effect of Valsalva maneuver, defecation [16, 35], or vaginal delivery [36], and also to evaluate the pelvic floor disorders [37].

3.1 Numerical Simulation—Generation of the Geometrical Models

To better understand the mechanical properties of the PFM with different pathologies, the first step is to accurately reproduce the female pelvic anatomy—the *levator ani* and surrounding structures (coccyx, obturator muscles, and *symphysis pubis*). In this sense, Janda et al. [35] and Parente et al. [37] built geometrical models using information obtained from cadaver measurements that correspond to one embalmed 72-year -old female cadaver. Parente et al. [37] and Silva et al. [38] simulated the vaginal delivery and modeling the fetus head using this geometric model of the PFM but Noakes et al. [16], Silva et al. [21], Saleme et al. [17], and Da Roza et al. [39] simulated the Valsalva maneuver and voluntary contraction using geometrical models obtained through the segmentation of MR images acquired in real subjects in the axial plane. Figure 2 shows the result of segmentation of axial MR images of the pelvis in an asymptomatic woman. Figure 2a shows the 3D solid model of the pubovisceral muscle after semiautomatic segmentation on an image analysis-processing software (using the Mimics® software). In Fig. 2b, the finite element mesh created in the Abaqus® software is presented after importing the 3D solid model presented in a).

To simulate different conditions—Valsalva maneuver, muscle contraction, or vaginal delivery—the boundary conditions to the PFM must be imposed to incorporate the existence of the surrounding structures, by visualizing muscle insertions in the MR images. The nodes corresponding to the insertion of the PVM in the different structures may be considered as fixed or with some degree of movement. Previous authors consider zero-displacement nodes in the points of

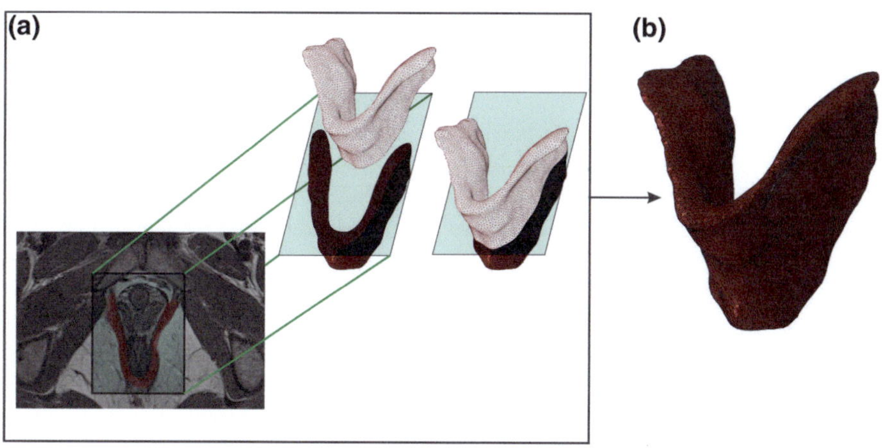

Fig. 2 The consecutive Magnetic Resonance axial images were segmented using an image analysis-processing software to create solid model (**a**), which is converted into finite element mesh (**b**)

insertion between PVM and the *symphysis pubis* and also the most superior portion of the PVM corresponding to the attachment to the *arcus tendineous levator ani* in the medial surface of the internal obturator muscle which constrains lateral movement [39]. In the boundary conditions of the coccyx, supero-inferior and medial-distal movements are allowed to simulate physiologic angular coccygeal movements [39].

To simulate the passive behavior of the PFM, the IAP has been measured at rest and the average of approximately 0.5 kPa [16] and has also been measured under different conditions during Valsalva maneuver, with average values of 4.5 kPa for subjects lying supine [16]. Accordingly, to simulate the Valsalva maneuver, Noakes et al. [16] and Silva et al. [21] applied a pressure of 4 kPa in inner surface of the muscle, since the computational models are based on images taken of the subject in a supine position and in a resting, the muscles already incorporate resting tone.

Usually, the PFM have a passive nonlinear behavior, but due to their skeletal muscle structure, namely the orientation of the muscle fibers can assume anisotropic behavior. Therefore, these muscles are distinct from other soft tissues because of their capability of active contraction [40]. Thus, to simulate the active behavior, it is also important to obtain an approximation of the muscle fiber direction, which is an important parameter for the constitutive model [39].

3.2 Establishment of the Biomechanical Properties of the Pelvic Floor Muscles

The constitutive laws are of critical importance for the analysis of mechanical behavior of soft tissues, i.e., to understand the PFM mechanics. Due to their nonlinear behavior, it is important the use of hyperelastic constitutive models. However, these constitutive models depend on variables that must be obtained experimentally. Martins et al [41] suggest three hyperelastic models for the study of soft tissues of the pelvis (Humphrey, Martins, and Veronda-Westmann), but other hyperelastic models can also be used, such as the Mooney–Rivlin, Neo-Hookean, Ogden, and Yeoh models.

To simulate the PFM action, many studies used mechanical properties obtained from experimental data (uniaxial and biaxial tests), because the live-subject analysis was impossible to achieve due to ethical reasons. The majority of the studies have used the material properties of cardiac and fascial muscles [42] and human pelvic tissue was characterized using in vitro mechanical tests on fresh cadavers. Brandão et al. [23] and Da Silva-Filho et al. [43] used numerical models to understand the role and the mechanical behavior of the pelvic structures in the development of UI and POP, by applying the material properties obtained from experimental studies with both normal and pathological specimens. Da Roza et al. [39] studied the UI in athletes and non-athletes and adopted the modified form of the incompressible transversely isotropic hyperelastic model proposed earlier by Humphrey and Yin

[44], and retrieving similar values of displacement retrieved from dynamic MR images and numerical models. Also, Noakes et al. [16] found similarity between MR images and numerical model using a Mooney–Rivlin material law to simulate the Valsalva maneuver.

Accordingly, as previously mentioned in this chapter, the inverse FEA is important to determine in vivo biomechanical properties of the PFM in a noninvasive manner, avoiding more complex and invasive approaches.

The inverse FEA was used by Kauer et al. [45] for characterization in vivo of soft tissues but resorted to an intra-operatively during hysterectomies while Silva et al. obtained in vivo biomechanical properties during Valsalva maneuver in a noninvasive way. The inverse FEA analysis implemented by Silva et al. [21] allowed to obtain the material constants according to different constitutive models (Neo-Hookean, Mooney–Rivlin). A more detailed explanation is now given below.

The Neo-Hookean (Eq. 1) and Mooney–Rivlin (Eq. 2) constitutive models are characterized by:

$$W = c_1(I_1 - 3) \tag{1}$$

$$W = c_1(I_1 - 3) + c_2(I_2 - 3) \tag{2}$$

where W is the strain energy function and c_1 and c_2 are the material constants to be determined that have dimensions of stress; I_1 and I_2 are the strain invariants of the right Cauchy–Green tensor [41].

Neo-Hookean and Mooney–Rivlin constitutive models are used for modeling the small strain nonlinear behavior of incompressible hyperelastic materials. To verify the differences between these two constitutive models, Martins et al. [41] used the silicone rubber and the soft tissues to compare and concluded that the Neo-Hookean model exhibits worse performance than the Mooney–Rivlin, and is unable to capture nonlinearity. Therefore, the number of material constants has influence on the optimal solutions [41].

This inverse FEA methodology uses an optimization algorithm to achieve the lowest difference between the position of the muscle, and for that purpose, dynamic images should be acquired to enable comparison with the numerical simulation, whether in Valsalva maneuver or contraction. The comparison between two curves allows estimating the error measure (Eq. 3): one curve corresponds to the position of the muscle, e.g., in a mid-sagittal image in the numerical model (shown on the right side of the Fig. 3) for each iteration and another curve represents its position in the dynamic mid-sagittal image acquired at Valsalva maneuver (left side, Fig. 3). For each iteration, the orthogonal distance between two curves is calculated and updates the material constants through the computation of the error measure and using the stopping criteria (Error < 10 mm), as suggested by Silva et al. [21].

$$Error = \sum_{i=1}^{np} \|d_{MRI_i} - d_{FEA_i}\| \tag{3}$$

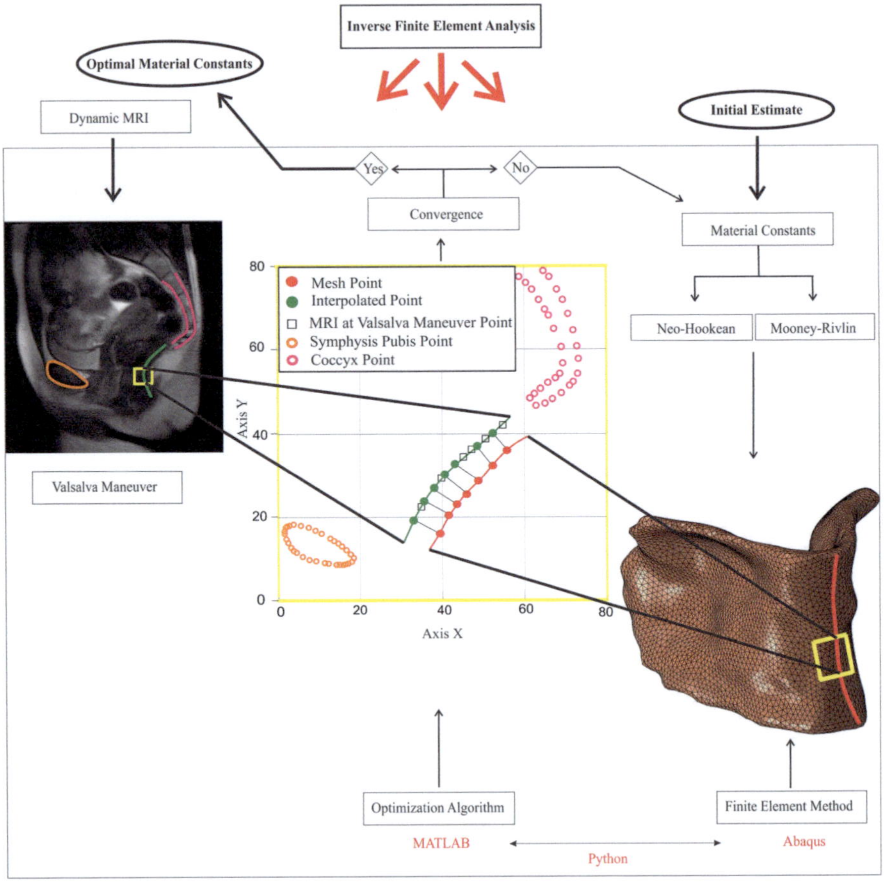

Fig. 3 Flow chart of the inverse FEA applied to obtain the material constants of the Neo-Hookean, Mooney–Rivlin models

where *np* is the number of points d_{MRI_i} used to define the first curve and d_{FEA_i} is the closest point projection of d_{MRI_i} on the second curve [21].

To illustrate this methodology with examples, three sets of conditions will be shown here: healthy muscles, a case of SUI, and another with POP. The three women had similar demographic characteristics and muscle morphology (see Table 1).

Through inverse FEA implemented by Silva et al. [21], the material constants of the Neo-Hookean and Mooney–Rivlin are higher for the muscles of the woman with prolapse (Table 2), i.e., the stiffness is 50 and 35% higher when compared with the ones of the woman without pathology, and 63 and 47% higher than the incontinent woman, respectively. On the other hand, the incontinent woman has an elasticity 27 and 18% lower than the one without pathology.

Table 1 Demographic characteristics and measurements of muscle thickness in asymptomatic woman, with SUI, and POP

Variable		Asymptomatic	SUI	POP
Age (y)		28	20	25
BMI (Kg/m^2)		20.57	20.66	23.97
Parity		0	0	0
Thickness [mm]	MidVa (left)	5.84	3.76	3.78
	MidVa (right)	6.44	5.19	4.54
	AC (left	5.11	4.47	1.90
	AC (right)	4.18	4.19	3.51

Table 2 Material constants for the pubovisceral muscle obtained for the Neo-Hookean and Mooney–Rivlin constitutive models obtained through inverse FEA

Variable [MPa]		Asymptomatic	SUI	POP
Neo-Hookean	c_1	0.030	0.022	0.060
Mooney–Rivlin	c_1	0.022	0.018	0.034
	c_2	0.008	0.005	0.025

There are few experimental studies about the PFM due to the orientation muscle fibers, but Martins et al. [34], analyzed the *levator ani* muscle obtained from female cadavers without any known pelvic trauma or pathology, and concluded that the tensile strength and tangent modulus are significantly smaller when compared with other pelvic soft tissues—bladder wall, vagina, or ligaments. However, Jean-Charles et al. [30] studied biomechanical properties of the vaginal tissue through uniaxial traction tests and concluded that women with prolapse are significantly more rigid than women without pathology (approximately, 89% for the material constant c_1 and 68% for the c_2). Lei et al. [32] studied the biomechanical behavior of vaginal tissue and also concluded that it was less elastic, and that stiffness increased in women with prolapse (the difference of the elastic modulus is approximately 15%), but also concluded that there is no significant difference between the different types of prolapse (moderate and severe types).

Previous authors affirmed that the SUI and POP can be associated with histological changes in the connective tissues—namely changes in the collagen that is the main structural component of the pelvic tissues [46, 47], but these changes are not visible in the MR imaging. In case of the SUI, Patel et al. [46] verified a significant reduction of type III collagen, which is not due to a decreased production of collagen but due to increased degradation of nascent collagen while Kerkhof et al. [47] showed that the women with POP have a decreasing total collagen content and increasing concentration of collagen type III. Therefore, the differences in the material constants can be associated with a histological response, as affirmed by Jean-Charles et al. [30] and Lei et al. [32].

4 Conclusions

In conclusion, the inverse FEA allowed establishing in vivo material constants of the female PFM for woman without pathology, with UI, and with POP, which substantiated more realistic results, and these constants are comparable with ex vivo experimental results [30, 32]. This methodology is completely noninvasive, using MR images to build the computational models, thus avoiding the use of ex vivo tissues collected postmortem, hysterectomy, or aspiration experiments during surgical proceedings [30, 32, 45], and in vitro material properties from experimental studies with both normal and pathological specimens [12].

Ex vivo studies of ligaments and endopelvic fascia have been concretized [2, 12, 14], through mechanical tests that allow studying their biomechanical behavior, such as the uniaxial and biaxial tensile tests, but there are few studies focusing the PFM in humans, due to the ethical challenges in obtaining samples and thus use animal models [15].

Due to limitations of the ex vivo studies on the PFM, the inverse FEA is an alternative to obtain the biomechanical properties and shows promising results for the subject-specific.

References

1. Schwertner-Tiepelmann N, Thakar R, Sultan AH et al (2012) Obstetric levator ani muscle injuries: current status. Ultrasound Obstet Gynecol 39(4):372–383. https://doi.org/10.1002/uog.11080
2. Abramowitch SD, Feola A, Jallah Z et al (2009) Tissue mechanics, animal models, and pelvic organ prolapse: a review. Eur J Obstet Gynecol Reprod Biol 144(1):S146–S158. https://doi.org/10.1016/j.ejogrb.2009.02.022
3. Tunn R, Goldammer K, Neymeyer J et al (2006) MRI morphology of the *levator ani* muscle, endopelvic fascia, and urethra in women with stress urinary incontinence. Eur J Obstet Gynecol Reprod Biol 126(2):239–245. https://doi.org/10.1016/j.ejogrb.2005.10.018
4. Moerman KM, Sprengers AM, Simms CK et al (2012) Validation of continuously tagged MRI for the measurement of dynamic 3D skeletal muscle tissue deformation. Med Phys 39 (4):1973–1810. https://doi.org/10.1118/1.3685579
5. Peng Q, Jones R, Shishido K et al (2007) Ultrasound evaluation of dynamic responses of female pelvic floor muscles. Ultrasound Med Biol 33(3):342–352. https://doi.org/10.1016/j.ultrasmedbio.2006.08.020
6. Yang JM, Yang SH, Yang SY et al (2009) Reliability of real-time ultrasound to detect pelvic floor muscle contraction in urinary incontinent women. J Urol 182(5):2392–2396. https://doi.org/10.1016/j.juro.2009.07.023
7. Mariappan YK, Glaser KJ, Ehman RL (2010) Magnetic resonance elastography: a review. Clin Anat 23(5):497–511. https://doi.org/10.1002/ca.21006
8. Gennisson JL, Deffieux T, Macé E et al (2010) Viscoelastic and anisotropic mechanical properties of *in vivo* muscle tissue assessed by supersonic shear imaging. Ultrasound Med Biol 36(5):789–801. https://doi.org/10.1016/j.ultrasmedbio.2010.02.013

9. Ringleb SI, Bensamoun SF, Chen Q et al (2007) Applications of magnetic resonance elastography to healthy and pathologic skeletal muscle. J Magn Reson Imaging 25(2):301–309
10. Jenkyn T, Ehman R, An KN (2003) Noninvasive muscle tension measurement using the novel technique of magnetic resonance elastography (MRE). J Biomech 36(12):1917–1921
11. Oudry J, Chen J, Glaser KJ et al (2009) Cross-validation of magnetic resonance elastography and ultrasound-based transient elastography: a preliminary phantom study. J Magn Reson Imaging 30(5):1145–1150. https://doi.org/10.1002/jmri.21929
12. Rivaux G, Rubod C, Dedet B et al (2013) Comparative analysis of pelvic ligaments: a biomechanics study. Int Urogynecol J 24(1):135–139. https://doi.org/10.1007/s00192-012-1861-5
13. Kirilova M, Stoytchev S, Pashkouleva D et al (2011) Experimental study of the mechanical properties of human abdominal fascia. Med Eng Phys 33(1):1–6. https://doi.org/10.1016/j.medengphy.2010.07.017
14. Jelen K, Herman H, Lopot F et al (2010) Women with incorrect pelvic floor statics: a biomechanical answer to the mechanical loading of the vagina-endopelvic fascia complex. Neuroendocrinol Lett 31(3):413–417
15. Rahn DD, Ruff MD, Brown SA et al (2008) Biomechanical properties of the vaginal wall: effect of pregnancy, elastic fiber deficiency, and pelvic organ prolapse. Am J Obstet Gynecol 198(5):590.e1–590.e6. https://doi.org/10.1016/j.ajog.2008.02.022
16. Noakes KF, Pullan AJ, Bissett IP et al (2008) Subject specific finite elasticity simulations of the pelvic floor. J Biomech 41(14):3060–3065. https://doi.org/10.1016/j.jbiomech.2008.06.037
17. Saleme CS et al (2011) An approach on determining the displacements of the pelvic floor during voluntary contraction using numerical simulation and MRI. Comput Methods Biomech Biomed Eng 14(4):365–370. https://doi.org/10.1080/10255842.2010.482045
18. Parente MPL, Jorge RM, Mascarenhas T et al (2008) Deformation of the pelvic floor muscles during a vaginal delivery. Int Urogynecol J Pelvic Floor Dysfunct 19(1):65–71
19. Brandão S, Parente M, Mascarenhas T et al (2015) Biomechanical study on the bladder neck and urethral positions: simulation of impairment of the pelvic ligaments. J Biomech 48 (2):217–223. https://doi.org/10.1016/j.jbiomech.2014.11.045
20. Martins P, Silva-Filho AL, Fonseca AM et al (2013) Strength of round and uterosacral ligaments: a biomechanical study. Arch Gynecol Obstet 287(2):313–318. https://doi.org/10.1007/s00404-012-2564-3
21. Silva M, Brandão S, Parente MP et al (2016) Establishing the biomechanical properties of the pelvic soft tissues through an inverse finite element analysis using magnetic resonance imaging. Proc Inst Mech Eng H 230(4):298–309. https://doi.org/10.1177/0954411916630571
22. Dietz HP, Shek C, Clarke B (2005) Biometry of the pubovisceral muscle and *levator hiatus* by three-dimensional pelvic floor ultrasound. Ultrasound Obstet Gynecol 25(6):580–585
23. Brandão S, Da Roza T, Mascarenhas T et al (2013) Moment of inertia as a means to evaluate the biomechanical impact of pelvic organ prolapse. Int J Urol 20(1):86–92. https://doi.org/10.1111/j.1442-2042.2012.03219.x
24. Chen B, Yeh J (2011) Alterations in connective tissue metabolism in stress incontinence and prolapse. J Urol 186(5):1768–1772. https://doi.org/10.1016/j.juro.2011.06.054
25. Yip C, Kwok E, Sassani F et al (2014) A biomechanical model to assess the contribution of pelvic musculature weakness to the development of stress urinary incontinence. Comput Methods Biomech Biomed Eng 17(2):163–176. https://doi.org/10.1080/10255842.2012.672564
26. Delancey JO, Miller JM, Kearney R et al (2007) Vaginal birth and de novo stress incontinence: relative contributions of urethral dysfunction and mobility. Obstet Gynecol 110 (2 Pt 1):354–362. https://doi.org/10.1097/01.AOG.0000270120.60522.55

27. Thompson JA, O'Sullivan P (2003) *Levator plate* movement during voluntary pelvic floor muscle contraction in subjects with incontinence and prolapse: a cross-sectional study and review. Int Urogynecol J Pelvic Floor Dysfunct 14:84–88
28. Tunn R, Paris S, Fischer W et al (1998) Static magnetic resonance imaging of the pelvic floor muscle morphology in women with stress urinary incontinence and pelvic prolapse. Neurourol Urodyn 17(6):579–589
29. DeLancey JO, Morgan DM, Fenner DE et al (2007) Comparison of *levator ani* muscle defects and function in women with and without pelvic organ prolapse. Obstet Gynecol 109(2 Pt 1):295–302. https://doi.org/10.1097/01.AOG.0000250901.57095.ba
30. Jean-Charles C, Rubod C, Brieu M et al (2010) Biomechanical properties of prolapsed or non-prolapsed vaginal tissue: impact on genital prolapse surgery. Int Urogynecol J 21 (12):1535–1538. https://doi.org/10.1007/s00192-010-1208-z
31. Rubod C, Boukerrou M, Brieu M et al (2008) Biomechanical properties of vaginal tissue: preliminary results. Int Urogynecol J Pelvic Floor Dysfunct 19(6):811–816. https://doi.org/10.1007/s00192-007-0533-3
32. Lei L, Song Y, Chen R (2007) Biomechanical properties of prolapsed vaginal tissue in pre- and postmenopausal women. Int Urogynecol J Pelvic Floor Dysfunct 18(6):603–607. https://doi.org/10.1007/s00192-006-0214-7
33. Barber MD, Bremer RE, Thor KB et al (2002) Innervation of the female *levator ani* muscles. Am J Obstet Gynecol 187(1):64–71
34. Martins PAL (2010) Experimental and numerical studies of soft biological tissues. PhD. Thesis. Faculty of Engineering, University of Porto, Portugal
35. Janda S (2006) Biomechanics of the pelvic floor musculature. PhD. Thesis. Technische Universiteit Delft, Netherlands
36. Li X, Kruger JA, Chung JH et al (2008) Modelling childbirth: comparing athlete and non-athlete pelvic floor mechanics. Med Image Comput Comput Assist Interv 11(Pt2):750–757
37. Parente M, Natal Jorge RM, Mascarenhas T et al (2009) The influence of the material properties on the biomechanical behavior of the pelvic floor muscles during vaginal delivery. J Biomech 42(9):1301–1306. https://doi.org/10.1016/j.jbiomech.2009.03.011
38. Silva MET, Oliveira DA, Roza TH et al (2015) Study on the influence of the fetus head molding on the biomechanical behavior of the pelvic floor muscles, during vaginal delivery. J Biomech 48(9):1600–1605. https://doi.org/10.1016/j.jbiomech.2015.02.032
39. Da Roza T, Brandão S, Oliveira D et al (2015) Football practice and urinary incontinence: Relation between morphology, function and biomechanics. J Biomech 48(9):1587–1592. https://doi.org/10.1016/j.jbiomech.2015.03.013
40. Martins JAC, Pires EB, Salvado R et al (1998) A numerical model of passive and active behavior of skeletal muscles. Comput Methods Appl Mech Eng 151(3–4):419–433. https://doi.org/10.1016/S0045-7825(97)00162-X
41. Martins P, Jorge RN, Ferreira A (2006) A comparative study of several material Models for prediction of hyperelastic properties: application to silicone-rubber and soft tissues. Strain 42:135–147. https://doi.org/10.1111/j.1475-1305.2006.00257.x
42. Martins JC, Pato MPM, Pires EB et al (2007) Finite element studies of the deformation of the pelvic floor. Ann N.Y Acad Sci 1101:316–334. https://doi.org/10.1196/annals.1389.019
43. Da Silva-Filho AL, Martins PA, Parente MP et al (2010) Translation of biomechanics research to urogynecology. Arch Gynecol Obstet 282(2):149–155. https://doi.org/10.1007/s00404-010-1396-2
44. Humphrey J, Yin F (1987) On constitutive relations and finite deformations of passive cardiac tissue: I. A pseudostrain-energy function. J Biomech Eng 109(4):298–304
45. Kauer M, Vuskovic V, Dual J et al (2002) Inverse finite element characterization of soft tissues. Med Image Anal 6(3):275–287

46. Patel PD, Amrute KV, Badlani GH (2007) Pelvic organ prolapse and stress urinary incontinence: a review of etiological factors. Indian J Urol 23(2):135–141. https://doi.org/10.4103/0970-1591.32064
47. Kerkhof MH, Hendriks L, Brölmann HA (2009) Changes in connective tissue in patients with pelvic organ prolapse-a review of the current literature. Int Urogynecol J Pelvic Floor Dysfunct 20(4):461–474. https://doi.org/10.1007/s00192-008-0737-1

Computational Analysis of Pelvic Floor Dysfunction

Aroj Bhattarai, Ralf Frotscher and Manfred Staat

Abstract Pelvic floor dysfunction (PFD) is characterized by the failure of the levator ani (LA) muscle to maintain the pelvic hiatus, resulting in the descent of the pelvic organs below the pubococcygeal line. This chapter adopts the modified Humphrey material model to consider the effect of the muscle fiber on passive stretching of the LA muscle. The deformation of the LA muscle subjected to intra-abdominal pressure during Valsalva maneuver is compared with the magnetic resonance imaging (MRI) examination of a nulliparous female. Numerical result shows that the fiber-based Humphrey model simulates the muscle behavior better than isotropic constitutive models. Greater posterior movement of the LA muscle widens the levator hiatus due to lack of support from the anococcygeal ligament and the perineal structure as a consequence of birth-related injury and aging. Old and multiparous females with uncontrolled urogenital and rectal hiatus tend to develop PFDs such as prolapse and incontinence.

Keywords Pelvic muscle · Muscle fibers · Passive stretching · Pelvic floor dysfunction

A. Bhattarai · R. Frotscher · M. Staat (✉)
Biomechanics Laboratory, Institute of Bioengineering, FH Aachen University of Applied Sciences, Heinrich-Mußmann-Str. 1, Jülich 52428, Germany
e-mail: m.staat@fh-aachen.de

A. Bhattarai
e-mail: bhattarai@fh-aachen.de

Present Address:
R. Frotscher
TWT GmbH Science and Innovation, Stuttgart, Germany

S. Brandão et al. (eds.), *Women's Health and Biomechanics*, Lecture Notes in Computational Vision and Biomechanics 29,
https://doi.org/10.1007/978-3-319-71574-2_17

Highlights

- Based on the ultrathin plastination technique a 3D biomechanical model of the levator ani (LA) muscle could be constructed with physiological cross-section based on fiber orientation.
- The passive stretching of the LA muscle based on the fiber matrix constitutive model can simulate its behavior much better than isotropic models.
- The damage in anococcygeal ligament and perineal structures causes involuntary widening of the *levator hiatus* which can lead to pelvic floor dysfunctions.

1 Introduction

The *levator ani* muscle supports the pelvic organs and provides urinary and fecal continence by relaxing and contracting effectively [1]. During pregnancy, supporting connective tissues become laxer, and the *levator ani* muscle fibers stretch and widen the way out for fetal descend [2], which normally regains its shape and function after delivery [3, 4]. However, significant denervation and injury of the pelvic floor musculature during childbirth also stresses the ligaments and soft vaginal hammock that also supports the pelvic organs [5], resulting in a range of discomforts such as local bleeding, irritation, and pain. Pelvic organ prolapse, incontinence, and sexual dysfunction are some of the pelvic floor dysfunctions (PFDs) that increase progressively with age and menopause [6]. About 40 and 50% of the world female population is estimated to suffer from stress urinary incontinence [7] and pelvic organ prolapse [8], respectively. Current studies describe high prevalence of stress urinary incontinence, also among young athletes [9, 10]. PFDs significantly affect women's quality of life from young to old ages and often result in the need for complex surgery, which is associated with economic burden. Although new surgical techniques are emerging, treatments without sufficient information on muscle morphology and anatomy often increase the risk of post-operative complications. Therefore, computational modeling using radiological images and finite element (FE) simulation is one of the best tools to improve the understanding of PFDs, to establish the efficacy of such surgical procedures, and to allow virtual clinical trials to eliminate potential risks before clinical deployment.

Recent modifications of the continuum constitutive models for skeletal muscles are mostly based on the Hill's three-element model. Skeletal muscle mechanics is described by three elements: the contractile element (CE) and two elastic spring elements, one in series (SE) and another one in parallel (PE) with the contractile element [11]. Furthermore, Humphrey and Yin accomplished major improvements of Hill's model by including the influence of the muscle fibers on muscle contraction. They split the pseudostrain energy function into densely distributed extensible fibers and an isotropic ground matrix [12]. Martins et al. [13] modified Humphrey's model and proposed a three-dimensional muscle model consistent with the one-dimensional model of the skeletal muscles model proposed by Hill [11] and

Zajac [14]. Several modifications and new developments of the skeletal muscle model were proposed; see, e.g. [15–17]. Recently, Böl and Reese decomposed the muscle constitutive laws into an active and a passive part and considered the presence of different fiber types within the muscle tissue [18].

The *levator ani* muscle is a fiber-reinforced skeletal muscle that lies between the pubic bone at the front and to the coccyx at the back. Numerous computational studies have been performed to analyze the behavior of the pelvic floor muscles under loading conditions ranging from Valsalva maneuver [19] to extreme deformations such as sports [10], different stages of labor [20, 21], and various fetal head shapes [22]. FE simulation has served as a tool to predict the susceptibility of any damage, either tear or denervation of the muscle fibers during desired conditions of intra-abdominal pressure (IAP). On this basis, the implementation of the modified Humphrey model to investigate the passive relaxation of the *levator ani* muscle without activating the muscle using FE simulation may be relevant.

This chapter will: (a) present a 3D computational model of the female *levator ani* muscle with a focus on the anatomy and fiber organization; (b) include an integrated discussion of the constitutive model based on the tissue constituents, fiber, and ground substance; and (c) emphasize the importance of the muscle fibers during increased IAP by comparing the passive relaxation of the *levator ani* muscle with the results of dynamic magnetic resonance imaging (MRI). We will also discuss the possibility of a *levator ani* dysfunction due to lost support from the anococcygeal ligament and perineal structures, and also highlight further studies regarding active contraction of the *levator ani* muscles and related dysfunctions due to localized damages in the muscle fibers.

2 Pelvic Muscle Model

2.1 *3D Geometrical Model of the* Levator Ani *Muscle*

The model as shown in Fig. 1a was reconstructed from a 70-year-old female cadaver specimen obtained from a coroner. The detailed methodology of creating the computer model from the plastinated slices has been previously described [23–25].

2.2 *NURBS-Based Geometry and Finite Element Mesh*

The geometrical model is well suited to map the pelvic floor anatomy. However, the reconstructed surfaces show distorted faces. FE simulations are known to fail or at least affect the convergence due to the distortion of the mesh under extreme deformations. The Rhino software[1] is used to repair the mesh/geometry by

[1]http://www.rhino3d.com/.

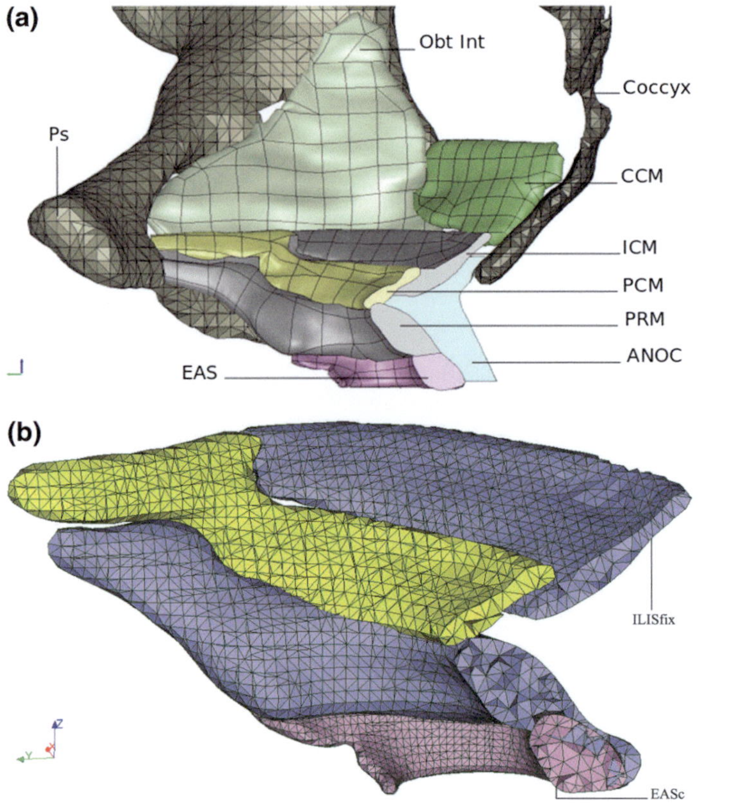

Fig. 1 **a** Sagittal section of the female pelvic muscle model created from the dataset of the plastinated reconstruction. ICM = iliococcygeus muscle, PCM = pubococcygeus muscle, PRM = puborectalis muscle, EAS = external anal sphincter, CCM = coccygeus muscle, ANOC = anococcygeal ligament, Ps = pubic symphysis, and Obt Int = obturator internus muscle. **b** A 3D finite element mesh of the *levator ani* muscle with two nodes EASc and ILISfix to compare muscle displacement and stress [20]

transforming into smooth freeform surfaces based on non-uniform rational B-splines (NURBS). The repaired geometry is then imported into the open source pre- and post-processor SALOME[2] to create an FE mesh. Considering the significant thickness of the *levator ani* muscle, the discretization of the volume should be performed. An example of a smooth FE mesh as shown in Figs. 1b and 2, which is constructed from 61538 linear tetrahedrons.

[2]http://www.salome-platform.org/.

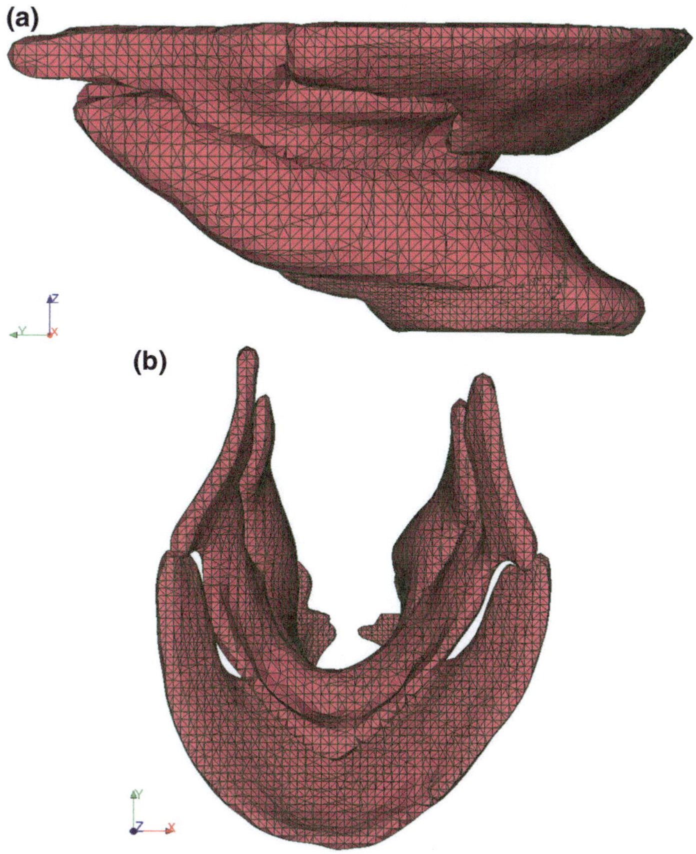

Fig. 2 FE mesh of the *levator ani* muscle in undeformed configuration. **a** lateral view: **b** superior view

3 Mechanical Modeling

3.1 Constitutive Model for Anisotropic Muscles

The majority of computer modeling studies investigating the *levator ani* muscle have assumed their mechanical behavior to be passive and isotropic. Hyperelastic constitutive relations, with either isotropic or anisotropic material properties, are often used to model large deformation of such tissues. Numerous early attempts have been made on cadaver dissections to describe the information about the fiber organization of the *levator ani* muscle, for which Janda et al. [26] were among the first to construct a useful model on the basis of the fiber arrangements. However, the development of the transversely isotropic constitutive models from Humphrey and

Yin and their successive improvements has put forward the novel idea of fiber contractility to evaluate the behavior of skeletal muscles.

The Helmholtz free energy in the model proposed by Humphrey and Yin for the skeletal muscle can be additively decomposed into

$$W_{muscle} = W_{incomp} + W_{matrix} + W_{fiber}, \tag{1}$$

with the incompressible (W_{incomp}), ground matrix (W_{matrix}), and fiber (W_{fiber}) contributions to the strain energy. Considering the muscle to be quasi-incompressible, the incompressibility contribution of the free energy takes the form

$$W_{incomp} = \frac{1}{D}(J-1)^2, \tag{2}$$

where D is the compressibility constant and the Jacobian $J = det\,\mathbf{F}$ is the volume change. $\mathbf{F}$ is the deformation gradient. In case of the perfectly incompressible muscle, $J = 1$. Since most of the soft tissues are regarded as perfectly incompressible, the free energy is equal to $\tilde{p}(J-1)$, where $\tilde{p}$ is a Lagrange multiplier, which has the physical meaning of hydrostatic pressure. The mechanical response of the matrix material is assumed to be isotropic; therefore, a hyperelastic model adopted by Humphrey and Yin [12] is used as

$$W_{matrix} = c\{exp[b(\bar{I}_1 - 3)] - 1\}, \tag{3}$$

where c and b are the material constants and $\bar{I}_1$ is the first principal invariant of the unimodular part of the right Cauchy–Green tensor $\boldsymbol{C} = \mathbf{F}^T\boldsymbol{F}$. It can be calculated as $\bar{I}_1 = tr\,\bar{\boldsymbol{C}} = tr\left[\bar{\boldsymbol{F}}^T\bar{\boldsymbol{F}}\right] = J^{-2/3}tr\,\boldsymbol{C}$, where $\bar{\boldsymbol{F}} = J^{-1/3}\boldsymbol{F}$ is the unimodular part of the deformation gradient. Further, the passive strain energy stored in the muscle fiber is

$$W_{passive} = T_0^M A\left\{exp\left[a\left(\bar{\lambda}_f - 1\right)^2\right] - 1\right\}\forall\bar{\lambda}_f \geq 1, \tag{4}$$

where A and a are the material constants, T_0^M is the peak stress when the muscle is fully activated, and $\bar{\lambda}_f$ is the fiber stretch ratio in the undeformed configuration. For $\bar{\lambda}_f < 1$, the strain energy is considered to be zero. The fiber stretch is given as,

$$\bar{\lambda}_f = \sqrt{\boldsymbol{N}^T\bar{\boldsymbol{C}}\boldsymbol{N}} = \sqrt{\bar{\boldsymbol{C}} : (\boldsymbol{N} \otimes \boldsymbol{N})}, \tag{5}$$

where $\boldsymbol{N}$ is the unit vector parallel to the preferred muscle fiber direction in the reference configuration which is directed anterior-posterior. Detailed calculation steps of the second Piola-Kirchhoff stress and the constitutive matrix are provided in Martins et al. [13].

3.2 Boundary Conditions

The anterior and posterior nodes of the *levator ani* muscle connected to the pubic symphysis and coccyx are completely fixed. Laterally, the *levator ani* is connected to the *obturator internus* (OI) muscle by the condensation of the fascia covering, commonly known as *arcus tendinous levator ani* (ATLA). Though, the OI muscle forms the pelvic sidewall, greater part of the OI muscle are unattached to the obturator foramen and ilium that allows 3D movement of the *levator ani* muscle along the attachment of the ATLA. Therefore, the lateral constraint has been released to induce free deformation of the *levator ani* muscle. Suddenly increased IAP acts over the pelvic organs, which is supported by the *levator ani* and suspended by connective tissues. This pressure is applied on the internal surface of the muscle.

3.3 Pressure Calculation

Efforts have been made to approximate the IAP during Valsalva maneuver and coughing for sitting and standing positions, see [19]. An average supine Valsalva pressure of 4 kPa has been averaged from the literature. In the present model, the area of the internal faces of the *levator ani* muscle that are connected to the pelvic organs via endopelvic fascia is summed up. Therefore, the IAP assumed in this numerical simulation was set equal to 3.1155 kPa.

4 Passive Stretching of *Levator Ani* Muscle

The material parameters of the Humphrey model (Eqs. (1–4)) are derived from Pato et al. [27] and are: $a = 12.43$, $b = 23.46$, $c = 3.79517355 \cdot 10^{-4}$ MPa, $A = 5.7270836 \cdot 10^{-4}$ MPa, $T_0^M = 6556.872 \cdot 10^{-4}$ MPa, and $D = 10^{-4}$ MPa. The finite deformation simulations for the computer model as shown in Fig. 2 were performed with the open source FE software, Code_Aster.[3] In the model presented in this chapter, the displacement of the *levator ani* muscle was compared with its initial position at rest, as shown in Fig. 3. As a reference, we used the vertical movement of the *levator* muscle of 27.9 mm set from a dynamic MRI acquisition in a nulliparous female during Valsalva maneuver from previously published work [19]. To verify the results, two points EASc and ILISfix on the FE mesh of the *levator ani* muscle are selected where larger deformation and stress were measured, see Fig. 1b. Figure 3a illustrates the vertical displacement of the *levator ani* muscle

[3]http://www.code-aster.org.

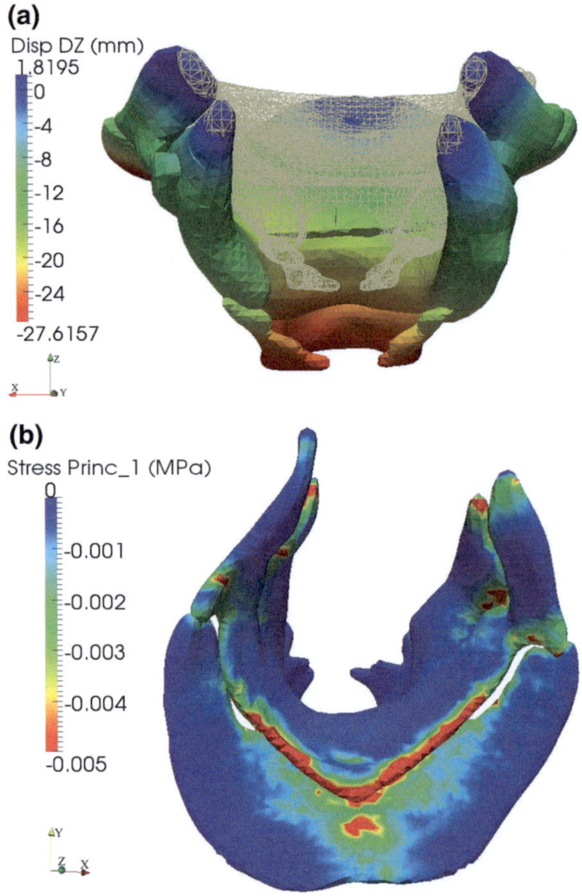

Fig. 3 FE simulation of the *levator ani* displacement due to an IAP. **a** Undeformed state (gray wireframe) and displaced *levator ani* muscle (solid): anterior view. **b** Stress distribution along the muscle. The larger principal stress in the simulation is, $\sigma_1^{max} = 3.27$ MPa: superior view

found in the FE simulation (25.5 mm), which is consistent with the aforesaid MRI reading [19].

Noakes et al. simplified the pelvic muscle as an isotropic hyperelastic material and used the Mooney–Rivlin material model. It offered a significant difference (13.7 mm) of the movement of the posterior region of the muscle, and for that reason, this chapter explicitly recommended adopting fiber-based constitutive models for improving the results. To this end, the Humphrey model has been implemented for the assessment of the influence of the fiber matrix on the displament under applied IAP. Despite the fiber being orientated toward the posterior direction, the *levator ani* muscle displaces about 1.64 times more than the one seen by Noakes et al. (7.7 mm vs. 4.7 mm, respectively). In both the Mooney–Rivlin and the Humphrey models, the justification for the differences might be the lack of any anterior–posterior support at the level of the EAS. Pelvic anatomy suggests that at this level, along with the posterior support from the fat tissues, the perineal body, the transverse *perinei superficialis*, skin, and the anal plate anchor the EAS muscle anteriorly and posteriorly. All of these structures are absent in the computational model. Thus, one can predict a case with such possible (extra) widening of the *levator hiatus*, which is due to the loss of structural integrity of the perineal structures and anal plate during vaginal delivery.

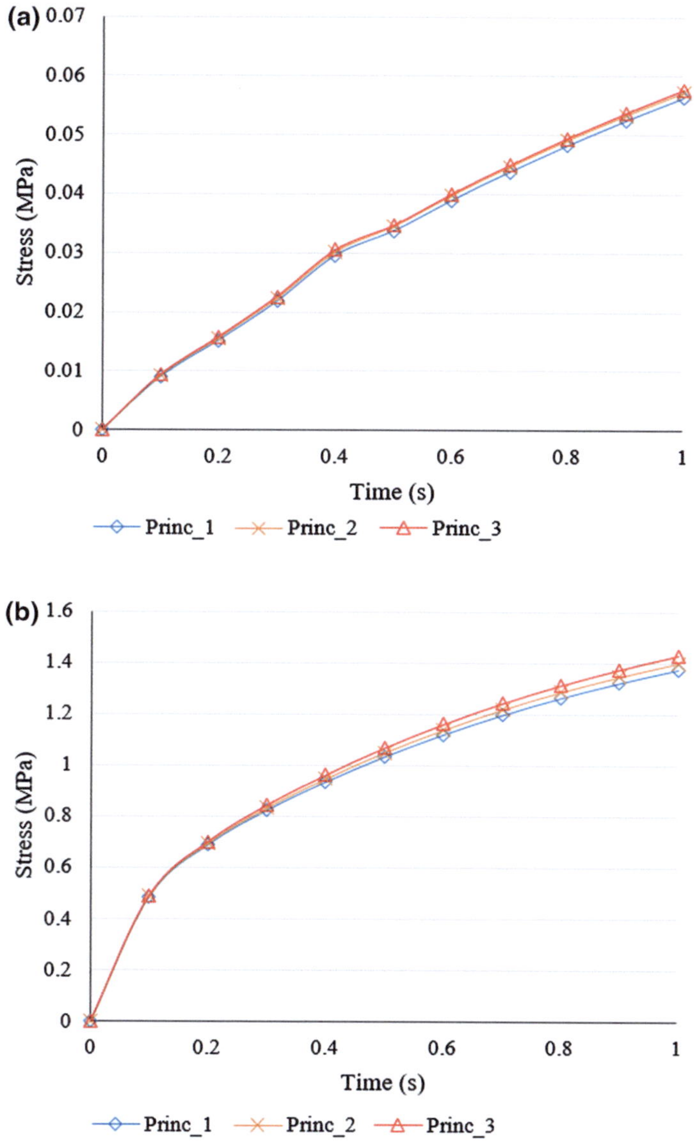

Fig. 4 Principal stress variation on nodes. **a** EASc: **b** ILISfix

Stress contours in the deformed pelvic muscle due to IAP are plotted in Fig. 4. The Cauchy stresses are compared at two selected nodes on the muscle, ILISfix (which is fixed to the coccyx) and EASc (which moves according to the deformation of the whole muscle), see Fig. 4. The stress value at the external anal sphincter, the exit of the anal canal, is quite low, however, around ILISfix, the part of the iliococcygeus muscle connected to the coccyx bone is highly stressed. Chances of localized damage of the *levator ani* are certainly high during extreme physical activities such as heavy physical exercise and straining during obstructed defecation.

5 Discussion

PFDs may be associated with impaired or injured muscles and connective tissues in the pelvic cavity due to overstretched pelvic floor by birth-related *levator ani* injury, chronic coughing, or obesity. Progressive weakening of the muscle, especially of *pubococcygeus* and its lateral connective tissue attachment to the vaginal walls fails to contract effectively to close the hiatal opening. As a result, vaginal walls move downward and are exposed to the differential pressure between abdominal and atmospheric pressure. This pressure difference further widens the *levator hiatus* and pulls the rest of the pelvic organs, which in turn, stress and stretch the supporting ligaments. Over time, due to excessive stretching, ligaments lengthen permanently and fail to support. Results are often seen as (urinary and fecal) incontinence and prolapse (cystocele, urethrocele, rectocele, and vaginal vault prolapse).

This chapter introduced some of the relevant physical and mechanical issues in the female *levator ani* muscle modeling which allowed studying its complex behavior during increased IAP. The results of the 3D *levator ani* muscle model developed from a subject-specific dataset of an old female cadaver qualitatively captured the passive relaxation of the pelvic muscle. The implemented Humphrey model is very promising to describe the passive mechanics of the female pelvic muscle due to IAP. Muscle fibers are qualitatively and quantitatively able to resist the exerted pressure and restrict the vertical *levator ani* muscle movement.

5.1 Limitations of the Model

Although the investigation of the passive response of the *levator ani* muscle has been done with great detail, its posterior movement was nevertheless larger than it has been observed in the reference MRI (7.7 vs. 4.7 mm), for which structural and mechanical reasons may have contributed. The presented model lacks some supporting structures such as perineal structures, fatty tissues, and anococcygeal ligament. Due to hormonal changes in old ages, weak support from such tissues and posterior shifting of the *levator ani* muscle are expected. Since the material parameters used in the simulation are adopted from the autonomously contracting heart tissue, differences in the behavior might be the other reason of different displacement measurement in *levator ani* muscle. The presented pelvic model is adopted from the cadaver *levator ani* muscle; lack of the muscle tone or the neural excitation in the mechanical model cannot sufficiently represent the in vivo muscle deformation. Thus, decreased muscle tone from an aged female leads to a decreased ability to balance the muscle contraction and the induced IAP. Nevertheless, results obtained from this simulation are additionally helpful to predict the location of regions where the muscle is critically stressed during increased IAP.

5.2 Pelvic Muscle Active Contraction

Clinically, PFD is considered when the relaxed pelvic muscle is not able to support and relocate the stable position of the pelvic organs. The presented model in this chapter considers only the passive stretching of the pelvic muscle fibers in the structural level. However, pelvic muscles are always excited to contract against the IAP which may be increased by physical activity. For this, active fiber contraction needs to be added into the model and simulated with proper neural excitation, which can be the next step forward in the modeling process. The strain energy describing the active contraction part is given by

$$W_{active} = T_0^M \left\{ \frac{1}{1000} exp\left[100\left(\bar{\lambda}_f - \lambda^{CE}\right)\right] - \frac{1}{10}\bar{\lambda}_f \right\} \forall \bar{\lambda}_f > \lambda^{CE}, \tag{6}$$

where λ^{CE} is the contractile stretch in series with an elastic element fulfilling the multiplicative split of the fiber stretch $\bar{\lambda}_f = \lambda^{CE}\lambda^{SE}$. The constant T_0^M is the maximum tension produced by the muscle at resting length $\left(\lambda^{CE} = 1\right)$. The constant T_0^M is the maximum tension produced by the muscle at resting $\left(\lambda^{CE} = 1\right)$. For the active muscle contraction, the parameter T_0^M is $6.82 \cdot 10^{-7}$ MPa. The contractile stretch can be calculated by equating the stress T^{SE} on the series elastic element and the stress T^{CE} on the contractile element adapted from Pato et al. [27] and Pandy et al. [28]:

$$T^{SE}\left(\lambda^{SE}, \lambda^{CE}\right) = T_0^M f^{SE}\left(\lambda^{SE}, \lambda^{CE}\right) \tag{7}$$

and

$$T^{CE}\left(\lambda^{CE}, \dot{\lambda}^{CE}, \alpha\right) = T_0^M f_L^{CE}\left(\lambda^{CE}\right) f_V^{CE}\left(\dot{\lambda}^{CE}\right)\alpha. \tag{8}$$

The functions f_L^{CE} and f_V^{CE} are the force–length and force–velocity properties of the muscle. The time-dependent muscle activation is caused by neural excitation with $\alpha \in [\alpha_{min}, 1]$ and $[\alpha_{min} \geqslant 0]$. For detailed derivation of the functions and the activation function, one can refer to Pato et al. [27].

6 Conclusions and Future Work

This chapter showed the relevance of an intact support from the anococcygeal ligament and perineal structures to maintain the diameters of the urogenital and rectal hiatus. PFDs such as incontinence and prolapse are often observed with widened hiatus that can be associated with permanent functional damage of the support system and reduced muscle tone due to a consequence of birth-related injury and aging. This chapter also explains that the passive stretching of the *levator ani* muscle is better modeled by using the fiber-based constitutive model developed by Humphrey and Yin, which gets closer to the MRI values than assuming isotropic behavior. This explains the emerging necessity of the anisotropic constitutive model for the skeletal pelvic muscle to enhance our ability to model and predict PFDs and vaginal delivery.

Further studies addressing the missing anatomic structures, correlated biomechanical properties between the healthy and the symptomatic females, activation of the pelvic muscle, and variations in the fiber contractility are required in the model to determine the absolute phenomena of the PFDs. An improved model avoiding such shortcomings can be used in the future to modify the muscle deformation results, which should be useful for clinical applications.

Acknowledgements

GEFÖRDERT VOM

Bundesministerium
für Bildung
und Forschung

The two first authors have been funded by the German Federal Ministry of Education and Research through the FHprofUnt project "BINGO", grant number 03FH073PX2.

References

1. Corton MM (2009) Anatomy of pelvic floor dysfunction. Obstet Gynecol Clin North Am 36 (3):401–419. https://doi.org/10.1016/j.ogc.2009.09.002
2. Downing SJ, Sherwood OD (1986) The physiological role of relaxin in the pregnant rat. IV. The influence of relaxin on cervical collagen and glycosaminoglycans. Endocrinology 118 (2):471–479. https://doi.org/10.1210/endo-118-2-471
3. Peschers UM, Schaer GN, DeLancey JO et al (1997) *Levator ani* muscle function before and after childbirth. Brit J Obstet Gynaec 104(9):1004–1008. https://doi.org/10.1111/j.1471-0528.1997.tb12057.x
4. Petros PEP, Ulmsten UI (1990) Pregnancy effects on the intravaginal sling operation. Acta Obstet Gynecol Scand 69(S153):77–78. https://doi.org/10.1111/j.1600-0412.1990.tb08038.x
5. Wall LL, Norton PA, DeLancey JOL (1993) Practical Urogynecology. Williams and Wilkins

6. Lin YH, Liu G, Li M et al (2010) Recovery of continence function following simulated birth trauma involves repair of muscle and nerves in the urethra in the female mouse. Eur Urol 57 (3):506–512. https://doi.org/10.1016/j.eururo.2009.03.020
7. Kane AR, Nager CW (2008) Midurethral sling for stress urinary incontinence. Clin Obstet and Gynecol 51(1):124–135. https://doi.org/10.1097/GRF.0b013e318161e687
8. Barber MD, Maher C (2013) Epidemiology and outcome assessment of pelvic organ prolapse. Int Urogynecol J 24(11):1783–1790. https://doi.org/10.1007/s00192-013-2169-9
9. Eliasson K, Larsson T, Mattsson E (2002) Prevalence of stress urinary incontinence in nulliparous elite trampolinists. Scand J Med Sci Sports 12(2):106–110. https://doi.org/10.1034/j.1600-0838.2002.120207.x
10. Da Roza T, Brandão S, Oliveira D et al (2015) Football practice and urinary incontinence: relation between morphology, function and biomechanics. J Biomech 48(9):1587–1592. https://doi.org/10.1016/j.jbiomech.2015.03.013
11. Hill AV (1922) The maximum work and mechanical efficiency of human muscles, and their most economical speed. J Physiol 56(1–2):19–41. https://doi.org/10.1113/jphysiol.1922.sp001989
12. Humphrey JD, Yin FC (1987) On constitutive relations and finite deformations of passive cardiac tissue: I. A pseudostrain-energy function. J Biomech Eng 109(4):298–304. https://doi.org/10.1115/1.3138684
13. Martins JAC, Pires EB, Salvado R et al (1998) A numerical model of passive and active behavior of skeletal muscles. Comput Methods Appl Mech Eng 151(3–4):419–433. https://doi.org/10.1016/S0045-7825(97)00162-X
14. Zajac FE (1989) Muscle and tendon: properties, models, scaling, and application to biomechanics and motor control. Crit Rev Biomed Eng 17(4):359–410
15. Yucesoy CA, Koopman BH, Huijing PA et al (2002) Three-dimensional finite element modeling of skeletal muscle using a two domain approach: linked fibre-matrix mesh model. J Biomech 35(9):1253–1262. https://doi.org/10.1016/S0021-9290(02)00069-6
16. Blemker SS, Delp SL (2005) Three-dimensional representation of complex muscle architectures and geometries. Ann Biomed Eng 33(5):661–673. https://doi.org/10.1007/s10439-005-7385-0
17. McLean SG, Su A, van der Bogert AJ (2003) Development and validation of a 3-d model to predict knee joint loading during dynamic movement. J Biomech Eng 125(6):864–874. https://doi.org/10.1115/1.1634282
18. Böl M, Reese S (2007) A new approach for the simulation of skeletal muscles using the tool of statistical mechanics. Mat.-wiss. u. Werkstofftech. 38(12):955–964. https://doi.org/10.1002/mawe.200700225
19. Noakes KF, Pullan AJ, Bissett IP et al (2008) Subject specific finite elasticity simulations of the pelvic floor. J Biomech 41(14):3060–3065. https://doi.org/10.1016/j.jbiomech.2008.06.037
20. Jing D, Ashton-Miller JA, DeLancey JOL (2012) A subject-specific anisotropic viscohyperelastic finite element model of female pelvic floor stress and strain during the second stage of labor. J Biomech 45(3):455–460. https://doi.org/10.1016/j.jbiomech.2011.12.002
21. Havelková L et al (2016) The effects of fetal head trajectory on stress distribution in levator ani during vaginal delivery. In: Jorge RN et al (eds) BioMedWomen. CRC Press, Boca Raton, pp 189–192
22. Yan X, Kruger JA, Nielsen PM et al (2015) Effects of fetal head shape variation on the second stage of labour. J Biomech 48(9):1593–1599. https://doi.org/10.1016/j.jbiomech.2015.02.062
23. Sora MC, Jilavu R, Matusz P (2012) Computer aided three-dimensional reconstruction and modeling of the pelvis, by using plastinated cross sections, as a powerful tool for morphological investigations. Surg Radiol Anat 34:731–736. https://doi.org/10.1007/s00276-011-0862-2

24. Feil P, Sora MC (2014) A 3D reconstruction model of the female pelvic floor by using plastinated cross sections. Austin J Anat 1(5):1022
25. Bhattarai A, Frotscher R, Sora M-C, Staat M (2014) A 3D finite element model of the female pelvic floor for the reconstruction of the urinary incontinence. In: Oñate E, Oliver J, Huerta A (eds) Proceedings of WCCM XI–ECCM V–ECFD VI, Barcelona, Spain, 20–25 July 2014, pp 923–934. https://doi.org/10.21269/6507
26. Janda S, van der Helm F, de Blok SB (2003) Measuring morphological parameters of the pelvic floor for finite element modelling purposes. J Biomech 36(6):749–757. https://doi.org/10.1016/S0021-9290(03)00008-3
27. Pato MPM, Areias P (2010) Active and passive behaviors of soft tissues: pelvic floor muscles. Int J Numer Meth Biomed Eng 26(6):667–680. https://doi.org/10.1002/cnm.1351
28. Pandy MG, Zajac FE, Sim E et al (1990) An optimal control model for maximum-height human jumping. J Biomech 23(12):1185–1198. https://doi.org/10.1016/0021-9290(90)90376-E

Conclusion

Improvements in social, mental, and clinical conditions are determinant for female health and well-being, and have been accomplished along of years in an integrated way. As demonstrated in this book, Women's Health includes several areas and, in this regard, a multidisciplinary collaboration is essential. Combining technology with medicine and engineering allows promoting preventive strategies, optimizing medical imaging to perform biomechanical analysis, and improving surgical strategies or rehabilitation protocols. However, the current literature demonstrates several aspects both clinical and (bio)mechanical have yet to be investigated.

According to World Health Organization (WHO), European women live longer and healthier lives. The continuous and hard research that has been shown in this book in issues such as continence promotion, delivery process, breast cancer, mental health, improving diagnostic accuracy in imaging techniques, and computational analysis methods has yielded an enormous wealth of information, making (when possible) an informed choice the best possible, although sometimes not the easiest. Along the aging process, some kind of dysfunction will appear in women's life, but fortunately, in the majority of the cases, they are treatable or at least manageable. Clinicians, physiotherapists, and biomechanical engineers should work in collaboration. We imagine biomechanical models demonstrating the effect of specific methods to improve the physical therapy or surgical techniques in a patient-specific condition.

Progress continues to be made in the awareness and promotion of Women's Health. We consider this book as a preliminary initiative to gather but mostly to promote collaborative investigation focusing on broad areas of female-related issues during their lives—young, adult, and older women—emphasizing how different factors and disciplines could influence prevention, diagnosis, and treatment.

With our warmest regards, the Editors (by alphabetic order).

Isabel Ramos, Sofia Brandão, Teresa Mascarenhas, and Thuane Da Roza.

S. Brandão et al. (eds.), *Women's Health and Biomechanics*,
Lecture Notes in Computational Vision and Biomechanics 29,
https://doi.org/10.1007/978-3-319-71574-2

MIX
Papier aus verantwortungsvollen Quellen
Paper from responsible sources
FSC® C105338

If you have any concerns about our products, you can contact us on
ProductSafety@springernature.com

In case Publisher is established outside the EU, the EU authorized representative is:
Springer Nature Customer Service Center GmbH
Europaplatz 3, 69115 Heidelberg, Germany

Printed by Libri Plureos GmbH
in Hamburg, Germany